Reading Bohr: Physics and Philosophy

Fundamental Theories of Physics

An International Book Series on The Fundamental Theories of Physics: Their Clarification, Development and Application

Volume 152

Reading Bohr: Physics and Philosophy

by

Arkady Plotnitsky

Purdue University,
West Lafayette, Indiana, USA

Springer

A C.I.P. Catalogue record for this book is available from the Library of Congress.

ISBN-10 1-4020-5253-7 (HB)
ISBN-13 978-1-4020-5253-8 (HB)
ISBN-10 1-4020-5254-5 (e-book)
ISBN-13 978-1-4020-5254-5 (e-book)

Published by Springer,
P.O. Box 17, 3300 AA Dordrecht, The Netherlands.

www.springer.com

Printed on acid-free paper

Table of Contents

Preface

This book is an exploration of the relationships between physics and philosophy in Niels Bohr's work, in quantum mechanics, and, finally, in physics itself, as, in Galileo's phrase, a "mathematical science of nature." It reassesses the place of Bohr's thought and writing both in the history of modern physics, from Galileo and Newton on, and equally in the history of modern philosophy. At the same time, the extension of the project undertaken by the book to quantum physics itself (rather than only Bohr's interpretation of quantum mechanics) and physics in general is crucial to the project. My title may also be read, by replacing the colon with a comma, as "reading Bohr, physics, and philosophy."

The main reasons for this expansion of the project's scope are as follows. While the relationships between physics and philosophy in Bohr's work have been considered in commentaries on Bohr, the implications of Bohr's work for the history of the relationships between physics and philosophy have not. I shall argue, however, that these implications are significant not only for our understanding of the history of quantum theory or physics in general but also for our assessment of the future of both, even if we finally want *to move beyond* Bohr and perhaps especially if we do. It is difficult *to leave Bohr behind* in considering quantum theory and its history. But in this case we can "move beyond" without "leaving behind," just as we moved beyond classical physics to relativity and quantum mechanics and then to quantum field theory without leaving anything behind. This is what the project of the book ultimately aims to accomplish, as it ends with quantum field theory in Chapter 5, and the relationships between physics and philosophy in Chapter 6, the final chapter of this study.

I shall pursue this project by means of close readings of some of Bohr's key works on his interpretation of quantum mechanics as complementarity. This approach is somewhat unorthodox in the fields of history and philosophy of quantum theory, even in studies specifically dedicated to Bohr. It has, however, several advantages not only, self-evidently, for understanding Bohr's work but also for understanding quantum theory and physics, and the relationships between them and philosophy beyond Bohr's work. First of all, it allows one to address with greater rigor and effectiveness the key questions at stake in the Bohr-Einstein confrontation and ongoing debates concerning quantum mechanics still shaped by this confrontation. One can mention such perpetual subjects as the double-slit and other "archetypal" quantum-mechanical experiments, the nature of quantum probability, and the experiment of A. Einstein, B. Podolsky, and N. Rosen, and J. S. Bell's and related theorems, some of which will be discussed in detail in the book. The approach also enables one to perceive and articulate more sharply than previously the key developments and transformations of Bohr's interpretation of quantum mechanics as complementarity. Most significant among them were those that occurred, first, under the impact of Bohr's debate with Einstein and, second, under the impact of the developments of quantum theory, both quantum mechanics itself and quantum electrodynamics and quantum field theory. The subject, especially the importance of the second factor just mentioned (some among more recent studies have discussed the first factor), has not been adequately addressed in the literature, to the considerable detriment of our understanding of the history of quantum physics. The book aims to fill this lacuna. As I said, one chapter of the book, Chapter 5, will be devoted to the relationships between quantum mechanics and quantum field theory, and the epistemological questions these relationships pose. The relationships among classical physics, relativity, and quantum mechanics will be addressed throughout the book, as they were throughout Bohr's work. A closer reading of Bohr shows that they are considered there in more depth and with greater significance than previously realized, and, thus, helps us to gain a greater insight into these relationships, crucial for physics and for our understanding of what physics is and of how it works.

Indeed, while this may be especially true in Bohr's case, in part given the proportion of physics contained in verbal formulations rather than mathematical formulas, or formal logical deductions that have dominated the foundational work on quantum mechanics, I would argue more generally that the role of reading in physics is more significant than is commonly acknowledged. Physics is also reading. It is the interpretation of texts, as well as of (and often jointly with) physical theories themselves, which may be especially true when dealing with quantum mechanics and its interpretation, but is also true throughout the history of classical physics or relativity. The history of quantum mechanics, from the work of founding figures to the most recent developments, certainly offers remarkable examples of both, both in general and specifically as concerns our encounters with Bohr's ideas. This history appears to be indissociable from interpreting Bohr's ideas, from reading Bohr.

The peculiarity of Bohr's writings is in part due to the peculiar nature of quantum physics, and of Bohr's interpretation and epistemology of it. In commenting on the difficulties involved in "Discussion with Einstein on Epistemological Problems in Atomic Physics," arguably his most definitive work on quantum epistemology, Bohr said: "Rereading these passages, I am deeply aware of the inefficiency of expression which must have made it very difficult to appreciate the trend of the argumentation aiming to bring out the essential ambiguity involved in reference to physical attributes of objects when dealing with phenomena where no sharp separation can be made between the objects themselves and their interaction with the measuring instruments." Such separation and, hence, the description of (the properties of) quantum objects and processes themselves (as opposed to certain effects of their interaction with measuring instruments upon the latter) are impossible in Bohr's interpretation. This impossibility expresses the essence of Bohr's epistemology. As Bohr also argues, however, this impossibility "provides room for new physical laws," and opens a space of new possibilities for physics and knowledge in general. An argument of this type is indeed not easy to make, especially to make efficiently.

On the other hand, some of the peculiarities in question are peculiarly Bohr's, especially insofar as Bohr's key terms, such as phenomena, individuality, atomicity, or complementarity, have unconventional and sometimes idiosyncratic meanings, which is often the case in dealing with philosophical terms and concepts. Bohr's writings appear to pose more substantial demands than customary in scientific texts as concerns paying special attention to particular formulations; carefully adhering to the particular meaning of his terms; understanding the philosophical (rather than only physical and mathematical) structure of his concepts; writing in different languages involved (Bohr wrote and thought on the subject in several languages) and translations between them; and so forth. These demands are not always met by Bohr's readers, which leads to significant misunderstandings of his arguments. Naturally, my point is not that one cannot disagree with Bohr's views or criticize his arguments, but the special conditions, often missed by Bohr's critics, that a meaningful reading or, if necessary, criticism would entail in his case.

The present book, nearly unavoidably, follows Bohr in its presentation of its subject. The approach does carry a potential benefit of opening the discussion to a broader readership, beyond those comprised by physicists and philosophers. On the other hand, the situation is complicated by the task, which I thought imperative, of retaining the rigor invariably found in Bohr's writings when dealing with quantum phenomena and quantum mechanics. (Bohr's excursions beyond quantum physics, even when using his concepts, such as complementarity, are, as he admits, speculative and less thorough.) Even though Bohr famously insisted that one should make one's presentation of what is fundamentally at stake in quantum physics available to a willing and open-minded layperson, his writings, even, and in some respects, especially, his philosophical writings, are not easy. While they do not always require technical knowledge of physics and mathematics (sometimes they do, even if implicitly, and at key points), they are not an easy reading and certainly do not conform to the genre of popular exposition. His writings are not inaccessible, but they are not always immediately accessible, and demand considerable effort on the part of any reader, not unlike philosophical works, such as those of Kant or Hegel, whose thought, as I shall discuss in the last chapter of this book, defines modern philosophy, the philosophical aspects of Bohr's (or Einstein's) work included. This study is also an attempt to negotiate this difficult balance between rigor and accessibility in presenting Bohr's writings, in reading Bohr.

The study addresses primarily Bohr's *interpretation* of quantum mechanics, and most especially the version developed in the wake of EPR's argument and finalized in "Discussion with Einstein," which refines Bohr's earlier versions of complementarity. Accordingly, most of my epistemological claims pertain to this interpretation rather than to the experimental data or mathematical formalism of quantum mechanics (if they can be seen as independent of an interpretation), or other interpretations of quantum mechanics, including those associated with "the Copenhagen interpretation." The latter rubric must be applied with great caution, given the differences between such interpretations and the thought of the different figures involved, even those who are considered, and consider themselves, close to Bohr (Heisenberg and Pauli, among them). These differences are much greater than it is usually argued and often outweigh the shared features, important as the latter may be. I would argue that, once considered in all of its aspects, Bohr's interpretation (in the present reading or "interpretation") is unique and, I would also argue, uniquely radical epistemologically. On several occasions, which I shall specify as I proceed, I shall advance arguments, both those arising from within Bohr's interpretation and relatively independent ones, that exceed the limits of Bohr's interpretation and lead to more general claims. They concern in particular the status of Bohr's interpretation as *an* interpretation, one among many possible interpretations, of quantum mechanics.

The project of the book could have been pursued on an even broader scale and via a more extensive textual engagement with Bohr's writings, in particular by extending this engagement to Bohr's works preceding his work on quantum mechanics, beginning at least with those on his 1913 theory of the hydrogen atom. Tempting as it may be (and was to the present author), such an extension would amount to an immensely long, nearly interminable investigation, even if one were to restrict oneself to Bohr's work. I ended up by making a virtue out of necessity and, while retaining the emphasis on reading, conceived of the project as a collection of *essays*, a genre defined by the lack of completion or the claim of completion. The approach inevitably entails certain losses, especially in Bohr's case, since nearly every paragraph (and often a single sentence) of his work on complementarity offers a rich source of possible commentary and a platform for further thinking about Bohr, physics, and philosophy. The Introduction and, to some degree, Chapter 1 are designed to offer an introduction to Bohr's key ideas, discussed in detail later in this study. In general, however, in accordance with the genre of the essay, each chapter may, in principle, be read independently, which also leads to some repetitions, although I tried to keep such repetitions minimal.

Bohr's writings may themselves be seen as conforming to the genre of the essay. Bohr has never written a book that would offer a sustained exposition of his interpretation (he, again, offered several) or of quantum mechanics and the phenomena in question in it as complementarity. At most, he published collections of his articles, essays, on the subject, even though he saw quantum mechanics as a complete theory (within its scope) and this completeness was a major theme of his incomplete, essay-like writings. On the other hand, quantum mechanics may well be, and in the ultimate version of Bohr's interpretation is, irreducibly incomplete, even within its own scope, insofar as it offers no description or even conception of the ultimate objects and processes it is concerned with and even appears to imply that such a description or conception is in principle impossible. It may lead us, yet again (one does not need quantum mechanics to do so), to ask whether the philosophy written in the "book" of nature, or in "the book of nature" that we write, in part in the language of mathematics, is indeed a book or a collection of essays. The latter appears to be rather more likely, at least to the present author. This is not necessarily a bad thing, although it would make the "dream of a final theory" in physics all but impossible, which may, however, not be so bad either.

It is worth stressing, however, that Bohr's essays offer us rigorous physics, as rigorous as any, and are sometimes compelled to pursue their arguments in an essay form in order to maintain this rigor. Planck's article introducing his black-body radiation law and with it quantum physics and Heisenberg's first paper of quantum mechanics have something of this quality as well, as does Bohr's so-called Como lecture, "The Quantum Postulate and the Recent Development of Atomic Theory," which introduced complementarity. All of these works may be seen as essays. Most of Bohr's endlessly revised writings were always essay-like, never finished. This was even how Bohr defined a "manuscript": as something to be further worked on. In all of these cases, however, his science was as rigorous as it could be, as was the quality of thought, coupled to a strength of conviction, which is, however, not the same as believing in delivering the final word on the subject. Only these qualities, always found in Bohr's works, define an essay, while this type of belief, never found in Bohr, is antithetical and inimical to it.

One can at most hope for, and certainly cannot count on, coming close to such works in undertaking the project of an essay. Going astray on such an adventure is more likely. All one can do is to try one's best to stay the course.

Acknowledgements

I am very grateful to Alwyn van der Merwe for inviting me to contribute to the series, *Fundamental Theories in Physics*, and Springer for publishing the book. I owe a large debt of gratitude to many mathematicians and physicists, whose ideas helped me in my work and with whom I had the privilege to discuss the project of this book. These exchanges proved to be invaluable in my work on the book and beyond it. I am especially grateful to N. David Mermin, whose work, ideas, and unerring critical judgment deeply affected all of my thinking about quantum physics. I would also like to thank Christopher A. Fuchs, Richard Gill, Tony Gonis, Kurt Gottfried, Gregg Jaeger, Basil Hiley, Andrei Khrennikov, Jan-Åke Larsson, Anthony J. Leggett, Peter Mittelstaedt, Asher Peres,† Rüdiger Schack, Marlan Scully, and Giuseppe Vitiello. I would like to thank my former professors and fellow students at the University of Leningrad, in particular Ludwig Faddeev, whose lectures and seminars on quantum mechanics and quantum field theory continue to have their impact on my work, including this book. I am grateful to Henry Folse for illuminating conversations about Bohr. Purdue University helped my work on the project by providing me with research and sabbatical leaves. I would like to thank those with whom I worked at Springer, in particular Kirsten Theunissen, Mieke van der Fluit, and Ian Mulvany, for their patience and professionalism. I am grateful to Colin Charlton for expert copyediting and virtuoso digital skills. I would also like to thank Nina, Marsha, Paula, and Inge-Vera.

An earlier version of Chapter 2 appeared, as "On the Precise Definition of Quantum Variables and the Relationships between Mathematics and Physics in Quantum Theory," in the special issue of *Foundation of Physics* (January 2006), dedicated to Asher Peres.

Introduction: Complementarity, Quantum Mechanics, and Interpretation

The aim of this introduction is to offer a brief outline of Bohr's complementarity as an interpretation of quantum mechanics and of the phenomena in question in it, quantum phenomena. This outline may be seen as primarily philosophical insofar as it aims to delineate the philosophical content of Bohr's key concepts. I shall, however, also discuss their physical content, both in its own right and in order to elucidate their philosophical content, or rather to use their physical and philosophical content to elucidate each other, and thus to understand better the reciprocity between physics and philosophy in the architecture of these concepts. I shall also address the status of complementarity as *an* interpretation, one among many possible interpretations, of quantum phenomena and quantum mechanics. First, I would like to establish more firmly the key general terms of my discussion here and throughout this study—quantum phenomena, quantum mechanics, and interpretation.

By quantum phenomena, I mean those physical phenomena in the analysis of which Planck's constant, h, cannot be treated as negligibly small. As will be seen, Bohr's special concepts of "phenomenon" and then "atomicity," crucial to the conceptual architecture of complementarity, especially in its ultimate version, were developed by him in order to develop a rigorous understanding and definition of such phenomena.

By "quantum mechanics" I mean the standard version of quantum mechanics (covered by Werner Heisenberg's or Erwin Schrödinger's formalism, or other, more or less mathematically equivalent, versions of the quantum-mechanical formalism, such as those of Paul Dirac or John von Neumann), rather than alternative accounts of the experimental data in question, such as Bohmian mechanics, for example.

By "interpretation," I mean primarily an explication of the physical content of a given physical theory, specifically of its mathematical formalism cum the experimental data to which this formalism relates by way of suitably idealized descriptions, predictions, and so forth. At the same time, an explication of physics often and, in a certain sense, always involves epistemological and otherwise philosophical considerations, which appear especially difficult to avoid in the case of quantum mechanics and which are manifestly significant for my argument in this study. Certain interpretive and epistemological adjustments are often necessary even when moving from different mathematical versions of quantum mechanics, although these versions may be seen as equivalent mathematically or in terms of their predictive capacities.

One of the main reasons for the significance of epistemological considerations for this study is that Bohr's complementarity defines the physical content of quantum mechanics in expressly epistemological terms and takes an epistemologically radical position concerning the nature of quantum phenomena and quantum mechanics. It sees quantum mechanics as a theory that deals only with the effects of the interactions between quantum objects and measuring instruments upon those instruments, and, moreover, in general, predicts such effects only in statistical rather than, as is the case in classical mechanics, deterministic terms. It is this view that grounds Bohr's concepts of phenomenon and atomicity, which he eventually developed to ground his interpretation more firmly. On this view, quantum mechanics is not assumed in any way to describe the behavior of quantum objects or the emergence of the effects in question (which is due to the quantum interaction between quantum objects and measuring instruments), nor even these effects themselves. Each such effect is described by means of *classical physics*, which, however, can neither describe their emergence nor, in contrast to quantum mechanics, predict their appearance, individually or collectively.

Accordingly, as I shall discuss in Chapter 1, quantum mechanics may be seen (with due caution and qualifications) as a form of information theory, rather than as a theory describing the behavior of its ultimate objects, such as electrons in atoms, in terms of spatial-temporal dynamics, in the way classical mechanics describes its objects, such as planets moving around the sun. Indeed, in Bohr's view, such a description is expressly prohibited or, in his words, "*in principle* excluded." Quantum mechanics predicts, in general probabilistically, the appearance of certain information on the basis of certain other information already available through the data obtained in the experiments already performed. The physical elements carrying the units of this information can be described in terms of classical physics and measured in classical bits. By contrast, neither the totality of these units and, hence, the essential "architecture" of this information nor the physical emergence of the elements in question can be described either by means of classical physics or by means of quantum mechanics, or conceivably by any means available to us. This statement explains my appeal to the radical character of this epistemology, meaning by "radical" that which is related to the fundamental root of the situation and transforms it in equally fundamental ways, in other words, something both fundamental and far reaching. Quantum mechanics, however, is able to predict the appearance of the individual and collective numerical data and informational configurations in question. Accordingly, the peculiar *architecture* of units or bits of classical information itself carries "information," which can be conveyed or transmitted (through the data, experimentally obtained or predicted by means of quantum mechanics) by classical, as well as by quantum, means but which can only be generated by quantum and never by classical means.

In this view, quantum mechanics *predicts* but does not *describe*: it predicts the appearance of certain observable and measurable effects and of certain configurations of these effects but does not describe the ultimate dynamics of their emergence. The physical elements of the configurations that it predicts and certain among their arrangements are describable and are, in this interpretation, described by means of classical physics. For example, one can use classical physics to describe "dots" on the screen or their different patterns (either the "interference" or "no-interference" pattern) in the double-slit experiments, at least as a suitable and, for the purposes of quantum-mechanical predictions, sufficient idealization since these "dots" are highly complex objects that appear as "dots" only at a low resolution. By contrast, the physical ("quantum") objects and processes responsible for the emergence of these elements and configurations are beyond any possible description.

As it follows recent developments in quantum information theory and adopts its language, the characterization just offered also indicates an *extension* or a particular inflected interpretation of Bohr's interpretation of quantum mechanics as complementarity. I shall explain the nature of this extension or this inflection presently. It may be useful, however, to briefly summarize, first, the key epistemological feature of Bohr's interpretation or of the present interpretation of Bohr's interpretation, as they will appear in this study.

The term "complementarity" originates in Bohr's argument that certain situations of measurement (such as those reflected by Heisenberg's uncertainty relations) are always mutually exclusive, and yet as each equally possible at any given point and as both necessary at different points for building a comprehensive theoretical framework accounting for the totality of the data in question. Bohr's choice of the term complementarity to describe the situation is, accordingly, idiosyncratic, since the term usually conveys that, rather than being mutually exclusive, the pictures in question complement each other as parts adding up to a whole, which is rigorously impossible in Bohr's definition.

This feature leads Bohr to the radical epistemology of complementarity, now understood, as it came to be in his work, in the sense of his overall interpretation of quantum mechanics and of quantum phenomena themselves. This epistemology ultimately entails a uniqueness of *each* situation of quantum-mechanical measurement. As will be discussed in Chapter 4, this epistemology also leads, or is correlative to, a peculiar character of quantum probability insofar as quantum mechanics becomes, in this interpretation, a probabilistic theory of individual events or phenomena rather than only a statistical theory of multiplicities of them. In contrast to classical statistical physics, probability and chance become irreducible in quantum mechanics, while a more detailed analysis of the constitution of such phenomena themselves, individually or collectively, is, to return to Bohr's language, "*in principle* excluded," thus making classical and quantum physics *fundamentally* different from each other. By the same token, this interpretation makes any given quantum-mechanical situation of measurement or prediction unique and unrepeatable, and, thus, incompatible with any other actual situation of measurement. Beyond its apparent inescapability for the overall comprehensive to account for the physical situation in question in quantum mechanics, the concept of complementarity as the mutual exclusivity of certain types of measurements remains crucial. It defines (for example, through Heisenberg's uncertainty relations) what specific predictions the theory can, or cannot, make, thus reflecting both quantum mechanics' capacities, which are tremendous, and its limitations (as concerns what can in principle be known), which are fundamental. It also leads to the radical epistemology in question, including as concerns the unique nature of each measurement or prediction, and to the peculiar character of probabilistic considerations involved.

Classical physics, specifically Newtonian mechanics, may be, and commonly is, seen as both *describing* the behavior of the objects it considers *and predicting* the outcomes of this behavior. By the term "object" I refer to the objects of classical physical theories, rather than those of nature itself, since classical physics deals with objects and models defined by properties that are abstracted or idealized from the properties of *natural objects* (whose other properties are disregarded) so as to make such models mathematically describable. Within its proper scope, however, Newtonian mechanics offers an excellent descriptive approximation of the behavior of natural objects and excellent predictions concerning this behavior. Accordingly, *within these limits*, it may be seen as *describing and predicting* the behavior of natural objects. This statement is of course not applicable in classical statistical mechanics (as concerns *description* of the behavior of the systems considered) or in chaos theory (in this case as concerns *prediction* of the behavior of the systems considered). The underlying dynamics considered by these theories may, however, be seen as subject to the same epistemological model.

By contrast, in quantum mechanics models of this type appear difficult and perhaps impossible to apply, and in the present interpretation such models are strictly inapplicable. This interpretation does not assign or makes it impossible to assign to quantum objects properties and behavior conceived on the model of classical mechanics (e.g., position, momentum, and so forth, even individually, rather than only jointly, which would be more immediately prohibited by the uncertainty relations) or of any other type, "quantum" (in whatever sense), "object" and "behavior," among them. As noted above, the experimental data itself in question is seen as rigorously unaccountable by classical physics either in terms of predicting the outcome of quantum experiments and by any theory, classical or other (quantum mechanics included), in terms of physically describing the *emergence* of these data. These data, as physical phenomena manifest in measuring instruments, along with the behavior of the instruments themselves, are seen as describable in terms of classical physics but as predictable only by means of quantum mechanics. By contrast the present interpretation theorizes "quantum objects," whose interactions with measuring instruments is responsible for the data in question, in such a way that no conceivable properties could be assigned them. As I shall explain presently, "quantum objects," thus conceived, are more properly seen as an idealization of certain entities in nature that interact with our measuring instruments and, by virtue of these interactions, lead to the appearance of the data in question. Such entities may also be macroscopic, as are, for example, the so-called Josephson's devices. Their character as quantum objects is, however, defined by their microscopic constitution, quantum in character, which may or may not be the ultimate underlying constitution of all nature. In any event, quantum mechanics offers only a limited, nonrelativistic theory of this constitution, which ultimately requires higher-level theories, such as quantum field theories. By the same token, in this interpretation the mathematical formalism of quantum mechanics *does not describe* the behavior of quantum objects anymore than does any classical or classical-like physics, but (in general, statistically) predicts the outcomes of possible experiments on the basis of the outcomes of experiments already performed and the data (classical in its physical character) obtained in them.

This situation does not of course prevent us from defining certain inaccessible objects as quantum and assigning to them an identity (e.g., electrons, photons, etc.) or from speaking of certain "properties" *associated* with them, such as mass or charge. This is now done in terms of particular correlated measurable effects such objects are responsible for. In other words, these properties are those of certain parts of measuring instruments, which is why, following Bohr, I speak of such properties as *associated* with quantum objects rather than as properties *of* quantum objects. These properties, however, emerge by virtue of the interactions between these instruments and quantum objects (or again, something in nature, idealized as quantum objects) in certain specifiable and properly correlated situations of measurement. These circumstances entail a *rigorous inapplicability* of classical-like models, along with an equally *rigorous applicability* of the probabilistic considerations to the outcome of the relevant experiments and thus give rise to a new conception of chance and probability in physics.

One might see the conceptuality of classical physics as a particular, suitably refined, form of what we can in principle conceive of. As both Bohr and Heisenberg emphasized, classical physics may be seen as a refinement of our common perception and thinking, specifically as regards such ideas as location in space or time, motion, force, and so forth. This refinement, however, and conceivably any refinement of our mental capacities, may not reach the "objects" in question in quantum mechanics. Bohr's interpretation expressly places quantum objects beyond the reach of our means of conception, representation, knowledge, access, and so forth. The concept of "object," however we can conceive of it, becomes ultimately inapplicable as well, and the quotation marks around this term or any other term, for example, "quantum," referring to quantum objects, are presupposed throughout this book.

It also follows, however, that such theories can only approach these objects through their effects on the classical world. For, it is only on the basis of such effects that one may construct such objects in rigorous theoretical terms, rather than merely imagine them. The physical constitution of these effects is physically, conceptually, and phenomenally classical. Their emergence and overall informational architecture (such as that found in the double-slit experiments, the EPR-type correlations, and so forth), due to quantum objects in their interaction with our measuring instruments, are, as I said, beyond the reach of the classical theories. Thus, as Bohr argues throughout, classical physical concepts appear to be necessary and irreducible within certain limits, which we may call classical in turn. These concepts, however, have rigorous limitations when we use them in handling the key quantum-mechanical effects, for example and in particular, in view of the mutual exclusivity (complementarity) of the simultaneous usage of some of them, say, those of position or momentum, which must, at least in principle, be jointly determinable at any given point in classical physics. Furthermore, these concepts are strictly inapplicable to describing quantum systems themselves and their behavior. This inapplicability does not mean that certain specifically quantum (i.e., not found in classical physics) features, such as "spin," cannot be introduced—quite the contrary. The question is to what degree, if any, we can conceptualize these features, for example, "spin," at the quantum level in terms of classical (that is to say, any) concepts, as opposed to defining the field of measurable effects associated with them and developing a mathematical formalism for predicting such effects. Both of these we can do rigorously. As will be seen, if anything, "spin," a famously inconceivable "angular momentum" (a useful metaphor borrowed from classical physics but ultimately inadequate to describe "spin") is a good paradigmatic case of this situation.

In a possible contrast to Bohr's view, the analysis of these circumstances and thus of quantum phenomena and quantum mechanics just sketched and to be developed in this study views Bohr's complementarity as *an* interpretation, one among possible interpretations, rather than a definitive interpretation, *the* interpretation, of quantum mechanics or quantum phenomena. Bohr's position on this issue appears to be somewhat ambivalent, and certain of Bohr's statements appear to suggest stronger claims. By virtue of this ambivalence, or in general, this position is itself subject to interpretation and thus part of one's interpretation of Bohr's interpretation, a predicament that one cannot avoid, although, as I explained in the preface, reading Bohr's work may present more interpretive complexities than usual.

The present view of complementarity as *an*, rather than *the*, interpretation of quantum phenomena and quantum mechanics has significant epistemological consequences of its own. Arguably most important among them is that the inconceivability of quantum objects and processes is seen as an idealization defining the objects of quantum mechanics in the particular interpretation adopted here, rather than as a definitive claim concerning the ultimate facts of nature or of our interactions with nature at the quantum level, with which Bohr's complementarity is particularly concerned. This idealization allows one to infer the existence of something in nature that manifests its existence in and is responsible for certain phenomena in the classical macro word (or what we in turn idealize in these terms) but is itself irreducibly beyond anything we can experience through our interaction with nature or beyond anything we can possibly conceive of. In the present view, however, this type of inconceivable entities must be seen as the ultimate objects of quantum mechanics in the particular interpretation adopted here and *not* as objects of nature. Hence, I speak of idealization. Whatever exists in nature that is responsible for the experimental data in question might, in this view, remain beyond even this idealization. It is, as it were, at a double remove or a double rupture from us, and may, *in relation to this particular idealization*, be viewed inconceivable even as inconceivable, in contrast to the ultimate objects of the theory, which are conceived as inconceivable. In general, however, it may also be something else, either something similar to the present view or something classical-like in character, or something different altogether. As such, this something may also be subject to alternative interpretations, either involving quantum mechanics or based on alternative theoretical accounts.

Bohr's complementarity makes no claim upon the ultimate constitution of nature itself, in the first place, by virtue of the fact that this constitution is placed beyond any possible knowledge and conception. Viewing complementarity as *an*, rather than *the*, interpretation of quantum mechanics, however, allows for different interpretations of quantum phenomena, different theories of the data in question, or different interpretations of quantum mechanics itself. Indeed, the epistemology of complementarity as understood here may not apply in the case of higher-level quantum theories. My argument itself only applies to quantum mechanics as a theory operative within its particular scope and limits, just as classical physics is operative within its scope and limits, or various quantum field theories are within their respective scopes and limits. The epistemology of quantum field theory, beginning with quantum electrodynamics, is a separate and complex issue, which I shall address in Chapter 5. It may require still more radical renunciations of our epistemological ideas and ideals, as, as will be seen, Bohr has pointed out on several occasions. But then, it may not, and the epistemological prospects are even less certain for more comprehensive theories that are necessary (since our theories at present are manifestly incomplete) but yet to be developed, inevitably "beyond Bohr."

Chapter 1. Complementarity, Epistemology, and Quantum Mechanics as an Information Theory

1. THE NO-CONTINUUM HYPOTHESIS

The argument of this chapter extends primarily from Niels Bohr's and Werner Heisenberg's work.[1] This argument, however and to some degree the argument of this study as a whole also comprise a meditation on John Archibald Wheeler's "no" to "continuum" in his quantum-information-theoretical manifesto, "Information, Physics, Quantum: The Search for Links":

> No Continuum. No continuum in mathematics and therefore no continuum in physics. [...] Nothing so much distinguishes physics as conceived today from mathematics as the difference between the continuous character of the one and the discrete character of the other. Nothing does so much to extinguish this gap as the elementary quantum phenomena "brought to a close," as Bohr puts it, by "an irreversible act of amplification," such as the click of a photodetector or the blackening of a grain of a photographic emulsion. [...] [C]ontinuum-based physics, no; information [bit] based physics, yes. (Wheeler 1990, pp. 9-10)[2]

This type of "no-continuum postulate" or, more cautiously, "no-continuum hypothesis" appears to be inherent in Bohr's complementarity and may be seen as the proper meaning of what he calls "the quantum postulate," his interpretation of Max Planck's discovery of "the quantum action" in 1900, which inaugurated quantum physics. Planck's discovery revealed that radiation, such as light, previously believed to be a continuous (wave-like) phenomenon in all circumstances, could, under certain

---

[1] Many of the works to be cited by this study are found in Niels Bohr, *The Philosophical Writings of Niels Bohr*, 4 vols. (Bohr 1987; Bohr 1998) and John Archibald Wheeler and Wojciech Hubert Zurek, eds., *Quantum Theory and Measurement* (Wheeler and Zurek 1983), which will be hereafter referred to as *PWNB* and *QTM*, respectively. The materials from the *Archive for the History of Quantum Physics (Interviews)* will be referred to as *AHQP*.

[2] By now quantum information theory has become a wide-ranging and highly developed field, with many theoretical achievements to its credit and major prospects for practical applications, most notably in quantum cryptography and computing. See (Fuchs 2001 and Fuchs 2003) for illuminating discussions, especially useful in the present context, and further references.

conditions, have a discontinuous, quantum character. Bohr's work and the debate concerning quantum mechanics, including Bohr's confrontation with Einstein (which largely defined this debate), are concerned with the nature and meaning of this discontinuity. As will be discussed later in this chapter, Bohr ultimately redefined it as part of his new concept of atomicity, as against classical atomism, the idea of a limited divisibility of matter itself, extending from Democritus on. Accordingly, this redefinition is one of the primary concerns of this study as well.

The limit at which this discontinuity appears is defined by the frequency of the radiation and a universal constant of a very small magnitude, h, Planck's constant, which Planck himself termed "the quantum of action" and which turned out to be one of the most fundamental constants of all physics. The indivisible (energy) quantum of radiation in each case is the product of h and the frequency ν, $E = h\nu$. The role of Planck's constant h may be seen as analogous to the role of c in special relativity (the constancy of the speed of light in a vacuum in its independence from the speed of the source) in terms of both the necessity of a departure from classical theory and of introducing the first principles of a new theory. The rest, one might argue, follows relatively naturally in both cases, although less so in quantum theory, in which case it has also taken longer to develop the consequences of Planck's assumption. The parallel formulas for energy, $E=mc^2$ (admittedly, a consequence rather than a postulate in special relativity theory) and $E = h\nu$, further amplify the parallel.[3] In any event, both quantum mechanics itself and Bohr's interpretation of it as complementarity were born from the necessity to give a proper physical and epistemological meaning to quantum discontinuity discovered by Planck. The no-continuum postulate itself can be stated as follows:

> *Any observable phenomenon of quantum physics is either individual discrete (discontinuous) or is a discrete sum, indeed is a finite (if possibly very large) sum, of such individual phenomena or events; or, in the language of information, every (informational) record of a quantum-physical event is either that of an individual phenomenon or the sum of such records.*

The summing-up ultimately pertains to records of such events occurring over certain periods of time, as in the case of the collisions between quantum objects and the screen in the double-slit experiment. It follows that there are no continuous, such as wave-like, quantum phenomena. Certain composite classical phenomena, defined by the corresponding effects of the interaction between quantum objects and measuring

---

[3] Cf., on the other hand, Christopher A. Fuchs's argument in (Fuchs 2001, pp. 40-43), to which my statement in part responds. While Fuchs's program of re-deriving quantum mechanics from certain more natural quantum-informational postulates may prove to be viable, the differences in question between special relativity and quantum mechanics may not be as significant as Fuchs argues.

instruments upon the latter, are wave-like (i.e., we can speak of such wave-like features as "diffraction," "interference," and so forth), while the individual phenomena comprising them are particle-like. That is, these phenomena are *analogous* (but not identical) to those observed in the interactions between particle-like objects and measuring instruments in classical physics. The wave-like effects in question appear under certain specified experimental conditions, once a sufficiently large number of events are accumulated, say, once a large number of quantum objects pass through the slits in the double-slit experiment (to be discussed below) and once there are no devices which allow us to know, even in principle, through which slit each object passes. (If we could have such knowledge, the interference pattern would inevitably disappear.)

I follow Bohr, indeed later, post-EPR Bohr, in presenting the non-continuum postulate in terms of the effects of the interaction between quantum objects and measuring instruments upon those instruments rather than in terms of the quantum objects themselves and their properties. As indicated above, the appeal to such properties ultimately proved to be inadequate for a rigorous account of the situation, at least according to Bohr's view. The difference between the two views of the situation—that of seeing it in terms of properties of quantum objects and that of seeing it strictly in terms of certain effects of the interactions between quantum objects and measuring instruments—has defined the debate concerning quantum mechanics throughout its history. The formal statement defining the postulate would be the same in both cases. The difference is in the definition of phenomena involved.

In Bohr's view, at least his ultimate view, all available quantum phenomena are defined strictly in terms of certain, sometimes correlated, recorded effects, "practically irreversible amplification effects," such as the click of a photo-detector or the blackening of a grain of a photographic emulsion, rather than in terms of properties of quantum objects themselves (*PWNB* 2, p. 51). The assignment of such properties is prohibited in view of "the *impossibility of any sharp separation between the behavior of atomic objects and the interaction with the measuring instruments which serve to define the conditions under which the [observable] phenomena [in question in quantum physics] appear* (*PWNB* 2, pp. 39-40). Quantum characteristics—such as discreteness, discontinuity, individuality, atomicity (indivisibility), and so forth—are transferred to the level of phenomena or effects in this sense, eventually (sometime in 1940s) also leading Bohr to as new concept of atomicity. This transfer requires a terminological adjustment, insofar as the application of such terms becomes conceptual rather than strictly physical. That is, these terms now apply to certain physically complex entities, each involving the whole experimental arrangement, defined by certain effects of the processes taking place by virtue of such arrangements, rather than ever to single physical entities, whether quantum objects themselves or even point-like traces of physical events. A collision of a "particle" with a silver bromide screen is discrete (a "dot") only in a low resolution, since physically such a trace or the process that led to it is immensely complex.[4] To quantum objects themselves one can no longer ascribe any conceivable physical, either wave-like or particle-like properties, or indeed other properties, ultimately not even "quantum" or "objects," however these are conceived in specific terms.

It is often noted that the concept of "wave" has no physical significance in Bohr's interpretation. It is equally often forgotten, however, that, in Bohr's interpretation, the concept of "particle" is equally inapplicable at the level of quantum objects. In Bohr's interpretation, the ultimate objects in question in quantum mechanics allow us to see them neither in terms of waves nor in terms of particles, nor in terms of any other specifiable entities we can conceive of. As used by Bohr, the term "quantum object" refers to entities to which no specific properties or conceptions are applicable. This is in part why the wave-particle complementarity was never especially favored by Bohr, even though it is, arguably, the most famous and the most often invoked example of complementarity.

One might even argue that, in a certain sense, the idea of wave is more important for Bohr than that of particle, if we recall that in quantum mechanics the idea of wave takes a new significance with Max Born's interpretation of Schrödinger's wave function in terms of wave-like ("propagating") probabilities, with which one can map outcomes of possible future experiments, involving a given quantum object. It is true that this type of appeal to waves could only be symbolic or metaphorical, even apart from the fact that a verification of any such map or, as Erwin Schrödinger called it, such "expectation catalogue" is bound to involve multiple objects. For one has to repeat the whole experiment anew, *ab ovo*, with a new, identically prepared, quantum object in order to verify each of the predictions comprising the initial catalogue (Schrödinger 1935, *QTM*, pp.154, 158-159). (Moreover, the outcomes of the identically prepared experiments cannot be guaranteed to be identical, and usually are not, which makes quantum-mechanical predictions irreducibly probabilistic.) For in what sense other than symbolic or metaphorical could probabilities "propagate" or be like "waves"? On the other hand, such "waves" still give us a rigorous catalogue of probabilities in predicting the outcomes of the experiments involved. As such they become part of Bohr's interpretation of quantum mechanics as an irreducibly statistical theory, even (in contrast to classical physics) as concerns the outcomes of individual processes and events, a crucial aspect of complementarity that I shall discuss throughout this study especially in Chapter 4. By contrast, the role of the idea of particles in quantum mechanics is seen by Bohr as purely symbolic, when it is applied to quantum objects, and, throughout his

---

[4] Cf., the discussion of the situation in (Ulfbeck and Bohr 2001) and (Bohr, Mottelson, and Ulfbeck 2004), although the authors seem to me to misread Bohr in thinking that he subscribes to viewing quantum objects and processes themselves in terms of particles and their motions, a view that they rightly question.

works, especially his later works, he tends to speak, more neutrally, in terms of quantum objects, rather than particles.[5]

It is worth stressing that Bohr's concept of phenomenon or, correlatively, atomicity applies only to individual phenomena as outcomes of single experiments. Collective configurations must be seen as collectivities of individual phenomena in Bohr's sense. Thus, in the double-slit experiment, the configurations that do display an interference pattern vs. those that do not must now be seen as *collectivities* of distinct types of individual phenomena (events, effects, and so forth), rather than different "phenomena," if we use the term *in Bohr's sense*. Once formed, such collective configurations could be seen as single phenomena in other senses, such as that of Edmund Husserl's phenomenology, or as a classical physical object—a plate with "dots" on it.

By the same token, in this type of interpretation, the laws (of a fundamentally statistical nature) of quantum theory give order only to the collectivities of individual events or rather individual effects/phenomena that are the outcomes of such events. By contrast, as, among others, Wolfgang Pauli stressed, when considered by itself, each such event is, in general, not comprehended by law. At least it is not comprehended by law in the way it would be in classical mechanics, which is indeed defined by a (causal) comprehension and (realist) representation of individual processes and events that it considers. Thus interpreted, quantum mechanics only predicts the probabilities of and correlations between certain events rather than describes individual physical processes (pertaining to quantum objects) in space-time. This was one of Heisenberg's decisive new insights in introducing his matrix mechanics, the insight that, as will be seen presently, was further radicalized by Bohr to the point of the impossibility, in principle, of such a description or the analysis it would make possible, of making such an analysis "*in principle* excluded' (*PWNB* 2, p. 62). The character of quantum information is defined by these circumstances, which is perhaps also the greatest enigma of the quantum-mechanical view of nature, or at least of one such view, if not of nature itself. How does the order of the multiple (such that of the interference pattern of the double-slit experiment or of the correlations found in the EPR-type experiments) arise from the randomness of individual quantum events, if each such event is considered separately?

2. QUANTUM EPISTEMOLOGY AND QUANTUM INFORMATION

One of the basic postulates of information theory (classical or quantum) is that information can be treated like a measurable physical quantity, usually measured in digital bits. That a measurable physical quantity is also a form of information in the

---

[5] It is this point that is, I think, missed in both (Ulfbeck and Bohr 2001) and (Bohr, Mottelson, and Ulfbeck 2004).

sense of something that can be extracted or learned from nature in an experiment, recorded, communicated, and so forth has been around for much longer and has formed the experimental basis of modern physics since Galileo. This type of information is, in general, richer than that in question in information theory, since it involves such features as the physical meaning of such quantities, of messages involving them, and so forth, which are expressly outside the purview of information theory by virtue of the apparent impossibility of mathematizing such meanings. But how much richer is it, or how much poorer is the type of information considered in information theory in the case of quantum information? At the very least, it appears that the nature of the information in the sense of information theory involved in quantum measurement can help us to address the questions these other features pose, in particular for the epistemological foundations of quantum mechanics or, conceivably, all of quantum theory.

The question of the (apparently irreducible) difference between classical and quantum physics may be posed in terms of the difference in the nature and structure ("architecture") of the quantum vs. classical data as information. That is, is there a specific set of features of the quantum-mechanical data or information fundamentally pertaining to quantum mechanics and irreducibly distinguishing it from the data of classical physics? The answer, as we know, is positive (with further qualifications, especially concerning locality), and it has major implications for the possibilities of organizing and processing information itself, specifically in quantum cryptography and computing. This difference is now usually presented in terms of "quantum entanglement," following the argument of Einstein, Podolsky, and Rosen, David Bohm's reformulation of this argument in terms of spin, John S. Bell's and related theorems (such as the Kochen-Specker theorem), the experiments of Alain Aspect, and related developments. This difference, however, began to emerge already with Heisenberg's original paper on his (matrix) version of quantum mechanics. It may also be seen as defining Bohr's interpretation of quantum mechanics as complementarity, especially in its post-EPR version. In other words, the situation is not restricted to entangled quantum objects. On the other hand, as will be seen in Chapter 4, all quantum-mechanical predictions involve a de facto entanglement between the states of the quantum objects and the measuring instruments involved or, more accurately (this is what gets entangled with quantum objects), particular quantum strata of the constitution of these instruments. The concept of state requires further qualifications in this context, which I shall discuss to in Chapter 2.

The word information, as applied to quantum data, is found already in Bohr's early writing on complementarity, in particular in his important and, unfortunately, rarely discussed 1929 "Introductory Survey" to Atomic Theory and the Description of Nature (now *PWNB* 1). As the notes there:

> In fact, the indivisibility of the quantum of action [i.e., *h*] demands that, when any individual result of measurement is interpreted in terms of classical conceptions, a certain amount of latitude be allowed in our account of the mutual action between the objects and the means of observation. This implies that a subsequent measurement to a certain degree deprives the *information* given by a previous measurement of its significance for predicting the future course of the phenomena. Obviously these facts not only set a limit to the *extent* of the *information* obtainable by measurement, but they also set a limit to the *meaning* which we may attribute to such *information*. (*PWNB* 1, p. 18; emphasis added)

This is already close to quantum-informational language and thinking. The term, however, appears to be first used in the definition of complementary features of quantum-mechanical description in his "Discussions with Einstein on Epistemological Problems in Atomic Physics." The essay was originally published in 1949 in the "Schilpp volume," *Albert Einstein: Philosopher-Scientist*, just after the appearance of Claude Shannon's work on information theory (Schilpp 1949). While this conjunction is likely to be coincidental, Bohr's usage of the term here or in the passage just quoted (indeed both elaborations may be linked) may be given an information theoretical sense. According to Bohr, in contrast to classical physics, in quantum mechanics "evidence obtained under different experimental conditions cannot be comprehended within a single picture, but must be regarded as [mutually exclusive and] *complementary* in the sense that only the totality of the [observable] phenomena [produces the data that] exhausts the possible *information* about the [quantum] objects [themselves]" (*PWNB* 2, p. 40; emphasis added; also *PWNB* 3, p. 4).

"Produces the data," which I insert, appears to be a useful qualification, if one wants to give this claim an information-theoretical or otherwise quantitative content, since, as already indicated, the phenomena in question also have a particular qualitative character by virtue of the epistemology that they entail. Further qualifications are also necessary as concerns the phrase "information about the objects." For, by this point (in 1949), in Bohr's interpretation, or at least in the present interpretation of Bohr's interpretation, this information pertains only to the effects of the interactions between these objects (or what we infer as such) and the measuring instruments upon those instruments. This information tells us literally nothing about these objects themselves, except, which is of course crucial, that they exists and have a capacity to produce the particular effects in question and, indeed, that we are prevented from having any knowledge or even conception concerning their own nature. In classical physics it is always possible, at least in principle and in idealized cases, to combine such evidence (say, concerning the position and the momentum of a given object at a given time) by means of a single experimental arrangement or compatible arrangements to form a picture of the physical behavior of the objects themselves under investigation. By contrast, this is never possible, even in principle and in idealized cases, in quantum mechanics, at least in Bohr's interpretation. The main reason for this is that the experimental arrangements necessary to ascertain the value of each such variable will always be mutually exclusive, or *complementary* in Bohr's sense. Ultimately, any given quantum-mechanical situation of measurement is always unique and unrepeatable, singular, and as such is incompatible with any other actual situation of measurement. The mutual exclusivity in the sense of complementarity of, say, the position and the momentum measurements, leading to the uncertainty relations, remains crucial, however. For it reveals and makes irreducible "the *impossibility of any sharp separation between the behavior of atomic objects and the interaction with the measuring instruments which serve to define the conditions under which the [observable] phenomena [in question in quantum physics] appear*" (*PWNB* 2, pp. 39-40; Bohr's emphasis).

Accordingly, in this interpretation, the role of measuring instruments becomes constitutively irreducible in considering quantum phenomena and in quantum mechanics as a physical theory of these phenomena. As a result, the structure or architecture of information provided by the data in classical and quantum physics becomes irreducibly different. The "new kinematics" of Heisenberg's first paper on his matrix mechanics was already in effect defined in terms of information concerning the probabilistic effects of the interaction between quantum objects and measuring instruments upon the latter, rather than, as in classical physics, in terms of the motion of the objects in question. (Hence, it dealt with the "possible information" involved in Bohr's definition of complementarity just cited.) These effects and only they are also what could in principle be mathematized in quantum mechanics and then only in terms of predicting their quantitative aspects (in this interpretation).

Bohr's expression "possible information" is suggestive of the meaning of information as possible communicable data (for example, in classical bits) concerning quantum objects. While, however, the elements and specifically units (bits) of information are classical, it relates to and is the effect of something that cannot be comprehended within a single picture (such as those of classical physics), either at the level of quantum objects or even in terms of the effects of their interaction with measuring instruments upon those instruments. This is the defining "architectural" difference between classical and quantum information. Bohr's statement cited above can, thus, be seen as suggesting an "informational" interpretation of the uncertainty relations as correlative to their "complementarity" interpretation in terms of the mutually exclusive experimental arrangements necessary for determining and indeed defining the complementary variables involved, such as "position" and "momentum." As will be discussed in more detail later in this study, the statistical character of quantum mechanics and quantum information is an automatic consequence (*PWNB* 2, p. 34). In the statement cited above Bohr stresses a somewhat different aspect of *complementarity*, namely that of "completeness," specifically, again, as concerns the possible information, rather than that of the mutual exclusivity of certain situations of measurement, involved in the uncertainty relations. At the same time, in view of this mutual exclusivity, the completeness of Bohr's "complementary" acquires a peculiar, inherently nonclassical, nature. For, if the objects in question were those of classical physics, this type of data and information could not be seen as complete concerning those objects, the fact or at least a possibility that is especially bothered Einstein. In Bohr's and the present view, however, this information is as complete as it can possibly be under the circumstances of quantum mechanics, which makes the difference from the classical theories irreducible. In Bohr's words, "in quantum mechanics, we are not dealing with an arbitrary renunciation of a more detailed analysis of atomic phenomena," which would enable us to reinstate the classical ideal of completeness in the sense of describing the behavior of the objects in question, "but with a recognition that such an analysis is *in principle* excluded" (*PWNB* 2, p. 62; Bohr's emphasis). Any possible description of quantum objects and their behavior is ultimately "*in principle* excluded," rather than merely abandoned as a concern of quantum theory. The latter, more positivistic, view appears to transpire in Heisenberg's original paper on quantum mechanics and is, accordingly, radicalized by Bohr. All that is, and could possibly be, available to us is a particular type of information concerning the effects of the interaction between quantum objects and measuring instruments upon those instruments.

3. FROM HEISENBERG'S NEW KINEMATICS TO BOHR'S COMPLEMENTARITY

The argument just sketched extends primarily from the post-EPR version of complementarity and still more specifically, from the version of it presented in Bohr's "Discussion with Einstein," arguably the most definitive and comprehensive exposition of his ultimate views.[6] Bohr's post-EPR views, I argue, retreat from those of the famous Como version of 1927, "The Quantum Postulate and the Recent Development of Atomic Theory" (*PWNB* 1, pp. 26-51), which introduced complementarity, and return to Heisenberg's original ideas used in developing his matrix version of quantum mechanics. One of Bohr's most significant statements on the epistemology of quantum theory (most key epistemological propositions of the latter works may be seen as developing this statement) occurs in his 1925 survey "Atomic Theory and Mechanics," well before (two years!) he introduced complementarity in the Como lecture. It occurs before Erwin Schrödinger's wave mechanics, but immediately in the wake of Heisenberg's paper

---

[6] The best summation of Bohr's ultimate argument for complementarity appears to be offered in his 1958 short essay "Quantum Physics and Philosophy: Causality and Complementarity" (*PWMN* 3, pp. 1-7), which also contains valuable nuances, some of which I shall note throughout this study.

introducing his matrix mechanics, a "rational quantum mechanics," a "step," as Bohr rightly guessed, "probably of fundamental importance." Eventually it proved to be one of the most radical epistemological steps in the history of physics. Bohr writes: "*In contrast to ordinary mechanics, the new quantum mechanics does not deal with a space-time description of the motion of atomic particles* [a concept eventually abandoned altogether, along with waves]" (*PWNB* 1, p. 48; emphasis added).

Apart from Heisenberg's discovery of the uncertainty relations, which was crucial, the Como version is indebted most substantially to Louis de Broglie's and Schrödinger's "wave" theories (although not their interpretations of those theories), which by and large disappear from Bohr's radar by 1930, although Schrödinger's equation itself remains of course the defining equation of quantum mechanics itself. Bohr's 1935 reply to Einstein, Podolsky and Rosen was arguably the most decisive work in developing his ideas in the form here considered, although there is earlier evidence of this shift, around 1930, under the impact of his previous exchanges with Einstein. In the wake of the EPR argument, Bohr arrived at a more radical interpretation, to which Heisenberg's early work could have led him more directly, although this was not apparent at the time. This interpretation does not depend on either wave or particle theories or properties, not even in partial terms, in describing quantum objects, to whose characterization and behavior neither, or indeed any, description or theory is applicable. At the most, some properties of either theory are retained at the level of the effects of the quantum (and hence ultimately in turn indescribable) interaction between quantum objects and measuring instruments upon those instruments. Nor, accordingly, would one depend on the wave-particle complementarity (I said, never especially favored by Bohr) in developing this interpretation. Bohr might have been better off following Heisenberg all along, rather than taking a Schrödingerian detour. Bohr's initial commentary suggests that Heisenberg's "matrix mechanics" paper is *nearly* all one needs to develop a Bohrian epistemology of quantum mechanics.

I qualify this assessment because, first of all, this epistemology would and perhaps could not have been arrived at without a more developed quantum theory, possibly including quantum electrodynamics, from Paul Dirac's work on, although by that time Heisenberg's initial scheme nearly reached, with the help of Max Born and Pascual Jordan's work, a fully-fledged version of matrix mechanics. (Dirac's somewhat different version of quantum mechanics and his quantum electrodynamics, based on his new ideas, and Schrödinger's wave mechanics were yet to come as well.) Secondly, Bohr's exchanges with Einstein, especially those concerning the EPR and related arguments, brought Bohr's interpretation of quantum mechanics as complementarity to a new level. As indicated above, they moved Bohr beyond merely the view (which can be seen in more positivist, Machian, terms) that "in contrast to ordinary mechanics, the new quantum mechanics *does not deal* with a space-time description of the motion of atomic particles" (emphasis added). In his ultimate interpretation the possibility of offering such a description or any analysis of phenomena in question, beyond a certain point, is, according to Bohr, not merely arbitrarily renounced but is "*in principle* excluded." Bohr's post-EPR view is defined by this move beyond merely an *arbitrary renunciation* to an *unavoidable renunciation*, making "*in principle* excluded" anything that defines, or occurs with, quantum objects themselves from any possible analysis—physical, mathematical, or philosophical. Nevertheless, Heisenberg's original work on quantum mechanics clearly prepares the ground and brings one to the threshold of this even more radical epistemological position, perhaps the most radical hitherto in science and philosophy alike.

We recall that quantum mechanics, as a theory dealing with the motion of electrons in the hydrogen atom, was introduced in 1925-26 by Heisenberg and Schrödinger in two different ("matrix" and "wave") versions, and developed in the work of Born, Jordan, Dirac, Pauli, and (primarily in terms of interpretation) Bohr.[7] However, even as it has offered a degree of resolution to the problems posed by Planck's discovery (which the preceding, "old," quantum theory, developed primarily in the work of Planck, Einstein, Bohr, and Arnold Sommerfeld, failed to solve) and as it became in mathematical-theoretical terms the standard theory, quantum mechanics brought with it new epistemological complexities. Indeed, while it was expected to resolve the problems posed by the "old" theory along more classical lines, including as concerns both realism and causality, quantum mechanics actually extended some of these problems to their more radical limits, arguably, to the most radical epistemological limits physics encountered hitherto. As a result, it did not appear to some, Einstein and Schrödinger, among them, to offer a real solution to those problems and was seen by them as at best a provisional theory, even within its proper limits.

In particular, quantum mechanics appeared to be able only to predict, mostly statistically (it makes some exact predictions), the outcome of experiments in question, but was unable to describe the motion of quantum objects in the way classical physics would for classical objects. Nor did it predict in the same way either, since it gave chance an even more radical character by making it irreducible, both in practice and in principle, even in dealing with individual, rather than (as in classical statistical physics) only collective, behavior. The outcomes of collective behavior could in certain circumstances

[7] Most of the key papers are assembled in B. L. van der Waerden, ed., *Sources of Quantum Mechanics* (van der Waerden 1968). For the history of most developments discussed here, see J. Mehra and He Rechenberg's treatise, *The Historical Development of Quantum Theory* (Mehra and Rechenberg 2001), which contains an enormous wealth of factual material and remains an essential reference on the history of quantum mechanics. I should note, however, that, although the present study is indebted to Mehra and Rechenberg's work, it offers a different philosophical perspective on and a different type of argument concerning quantum mechanics and its interpretation, and of the work of the key figures involved, especially that of Bohr.

be subject to certain patterns or forms of order and (statistical) predictive law; the individual behavior is fundamentally or, in David Bohm's words, "irreducibly lawless" (Bohm 1995, p. 73). It may be more accurate to speak here of *sequential* behavior, since the ("ordered") collective effects in question would usually result from sequences of individual events, such as a large number of collisions between particles and a screen in the double-slit experiment, whereby a certain (interference-like) pattern would emerge.[8]

By the same token, all such manifest effects, individual or collective, appeared to pertain to certain parts of measuring instruments. This link was elevated into a postulate by Bohr and became a crucial feature of his interpretation, especially in his ultimate version of complementarity, according to which, as noted above, there is no other data one *could possibly account for*, rather than (along more positivist lines) the only data one *needs to account for*, in quantum physics. We can describe the impact, the physical effects, of quantum objects and processes upon our measuring instruments in a strictly classical—objective and realist—manner, which is an important point, often stressed by Bohr. We cannot, however, describe the ultimate nature of these processes themselves as responsible for these effects. In classical physics observable phenomena (now using the terms in its usual sense of what one observes, rather than that of Bohr) can be both properly *related* to and properly *correlated* with the observable properties of objects under investigation. In quantum mechanics, in this interpretation, observable phenomena and the data or information they provide can only be *correlated* with the behavior of quantum objects. It is not possible to ascertain the physical correlata of such correlations as properties of quantum objects or of their behavior, whether those of the quantum objects under observation or those of the quantum stratum of measuring instruments, through which the latter interact with quantum objects.[9]

Accordingly, it is hardly surprising that from the classical-like perspective, or according to classical-like expectations, one would not be satisfied with quantum mechanics, as a theory at the very least open to such an interpretation. The statistical nature of quantum predictions or the uncertainty relations are (correlative) experimental facts, rather than is features of the theory, although Einstein was inclined to see them in that latter way, which view in part grounded EPR's argument and related arguments by

---

[8] I am not referring to quantum statistics, which studies quantum multiplicities, just as classical statistical physics studies classical multiplicities, although quantum statistics is subject to the epistemology in question. Part of my argument is that, in the present interpretation, quantum phenomena entail and quantum mechanics offers the predictions that are, in general, statistical even as concerns individual processes and phenomena considered. Individual processes and phenomena considered by classical physics are covered, both descriptively and predictively, by classical mechanics as a (descriptively) causal and (predictively) deterministic theory. I shall return to the subject in Chapter 4.

[9] This point echoes N. David Mermin's (different) argument in (Mermin 1998a).

Einstein, and was one of the targets of Bohr's counterarguments. From another perspective, however, such as the one adopted here, one might say that quantum mechanics converted these problems, transpiring already in the old quantum theory, into ways of solution. By extending these problems to their more radical limits, some among the founders of quantum mechanics found its solutions in the deeper nature of these problems, albeit solutions that might require a radical epistemological position, unacceptable to Einstein and Schrödinger, or many others.

Heisenberg was first to adopt this strategy, characteristic of his intellectual temperament, by introducing his "new kinematics," introduced in his first paper on quantum mechanics and developed by him and others in a full-fledged quantum mechanics, the development powerfully presented by Heisenberg himself in his 1929 Chicago lectures, *The Physical Principles of the Quantum Theory* (Heisenberg 1930).[10] Traditionally, as the term "kinematics" indicates, it refers to a representation, usually by means of continuous functions, of the attributes of motion, such as positions (coordinates) or time, or velocities of a body. The representation of dynamic properties, such as momentum and energy, are dependent on and are functions of kinematical properties, but are also dependent on and are functions of the masses of the bodies involved. By contrast, Heisenberg's "new kinematics" referred its mathematical elements to what is observable in measuring instruments under the impact of quantum objects, rather than represented the attributes of these objects themselves. In addition, these elements were no longer functions, but infinite-dimensional square tables or matrices of complex variables with no classical-like, nor ultimately any relation, to the attributes of motion of quantum objects, but related only to the impact of the latter upon measuring instruments. These relations were established by means of certain, *ad hoc*, mathematical rules through which one could generate certain sets of real and indeed whole numbers (quantum numbers), corresponding to the results of the measurements in question. These rules are essentially equivalent to Born's square moduli rule for the wave function, a rule more general in nature, applicable to all quantum-mechanical predictions.

In his initial commentary, Bohr observed that the "fundamental importance" [of Heisenberg's step] was in "formulating the problems of the quantum theory in a novel way by which the difficulties [that besieged quantum theory since Planck] attached to the use of mechanical pictures may, it is hoped, be avoided" (*PWNB* 1, p. 48). Ultimately,

---

[10] This work summarizes the arguments of "[On] Quantum-Theoretical Re-Interpretation of Kinematical and Mechanical Relations" [*Über quantentheoretische Umdeutung kinematischer und mechanischer Beziehungen*] (van der Warden 1968, pp. 261-77) and "The Physical Content of Quantum Kinematics and Mechanics [*Über den anschaulichen Inhalt der quantentheoretischen Kinematik und Mechanik*]" (*QTM*, pp. 62-86) introducing, respectively, quantum mechanics and the uncertainty relations. The English translation of the second title is misleading and should read instead "on the representable (intuitable) content of quantum-theoretical kinematics and mechanics."

these difficulties were avoided by abandoning such a use altogether, at least in Bohr's interpretation. Remarkably, however, the formalism enables excellent statistical predictions concerning the outcome of experiments and the key physical laws, such conservation laws, could still apply, with proper adjustment (*PWNB* 1, p. 48). Bohr added:

> In [Heisenberg's] theory the attempt is made to transcribe every use of mechanical concepts in a way suited to the nature of the quantum theory, and such that in every stage of the computation only directly observable quantities enter. In contrast to ordinary mechanics, the new mechanics does not deal with a space-time description of the motion of atomic particles. It operates with manifolds of quantities which replace the harmonic oscillating components of the motion and symbolize the possibilities of transitions between stationary states in conformity with the correspondence principle. These quantities satisfy certain relations which take the place of the mechanical equations of motion and the quantization rules [of the old quantum theory]. (*PWNB* 1, p. 48)

Indeed the classical (Newtonian, in Heisenberg's paper, or more generally, Hamiltonian, as in most other approaches) equations of motion are formally retained in these relations, but are applied only to matrix variables and no longer describe the motion of quantum objects.

Bohr adopts the concept of new kinematics in the Como lecture, whose first section is entitled "Quantum of Action and Kinematics." Later in the lecture he says: "The new development [of quantum theory] was commenced in a fundamental paper by Heisenberg, where he succeeded in emancipating himself completely from the classical concept of motion by replacing from the very start the ordinary kinematical and mechanical quantities by symbols which refer directly to the individual processes demanded by [Planck's] quantum postulate (*PWNB* 1, pp. 70-71). The word "refer" need not mean "describe," and, I would argue, does not, even, in the Como lecture (let alone in Bohr's later works), at least as concerned the behavior of quantum objects between their interactions with measuring instruments. These "individual processes" do, however, generate what is actually observed in measuring instruments (each event involved being unique). All quantum-mechanical measurements now concerned only certain observable quantities pertaining to certain parts of measuring instruments in their interaction with quantum objects, rather than to quantum objects themselves. Accordingly, insofar as kinematical and dynamic properties of physical objects are involved at all, such properties are only those of certain parts of measuring instruments. Heisenberg considers atomic spectra, but this does not change the epistemology of the situation. His argument can be adjusted so as to refer to classical physical variables, such as position and momentum, pertaining to certain parts of measuring instruments under the impact of quantum objects. These considerations make clear why Heisenberg's approach is much more conducive to Bohr's view than Schrödinger's wave mechanics, which aimed at relating its formalism to the space-time (wave) processes at the quantum level. Schrödinger's equation itself can of course be interpreted in accordance with Bohr's view.

A qualification may be in order, concerning Heisenberg's famous emphasis on the "quantities, which in principle are observable," in other words, magnitudes related to individual quantum effects in Bohr's sense, which, rather than properties of quantum objects and of their behavior, become subject to his "new kinematics." His theory qua theory was *not* founded on such quantities, at least not only on such quantities, but engaged with a much greater complexity of both experimental observation and theory production alike, and of their relationships, in particular their irreducible mutual reciprocity.

This complexity transpires already in Heisenberg's famous, but not always carefully read, opening statement: "The present paper seeks to establish a basis for theoretical quantum mechanics founded exclusively upon *relationships* between quantities, which are *in principle* observable" (emphasis added). "Relationships" is the key word here, and the title of the paper was "[On] Quantum-Theoretical Re-Interpretation of Kinematic and Mechanical *Relations[hip]*" (emphasis added). "In principle" is crucial too, for, no matter how theory-laden and how complicated the processes of observation, the quantities in question could, in principle, be observed and, as it were, "kinematically" related to in the sense outlined above, while the classical-like physical (and perhaps ultimately any) properties of quantum objects and of their behavior could not. In other words, *dealing* with such, "in principle observable," magnitudes and founding a theory on the *relationships between them* is not the same as *founding the theory on them*, or *only* on them.

While working with the available data of quantum physics (such as the Rydberg-Ritz formulas and the Bohr frequency relations), Heisenberg's theory qua theory was *founded* above all on Bohr's correspondence principle and gave the latter a more precise meaning, as Bohr immediately grasped. As Bohr says in "Atomic Theory and Mechanics": "The whole apparatus of the quantum mechanics can be regarded as a precise formulation of the tendencies embodied in the correspondence principle" (*PWNB* 1, p. 49). The correspondence principle was used to argue that for large quantum numbers the data becomes the same as it would be in a classical case, at least as far as predictions are concerned (the description itself could no longer be the same). The principle was also used to argue, correlatively, that the equations should be formally the same as those of classical mechanics. From this viewpoint, Bohr's "correspondence principle" may have been his most original contribution (it is uniquely his) to quantum mechanics. Complementarity might be seen as a variation, quite original but a variation nonetheless, on Heisenberg's themes, in turn, however, depending on Bohr's correspondence principle.

This, as far as classical physics is concerned, "lethal" combination of the data and the correspondence principle leads to the remarkable features of quantum mechanics, both physical (Born's probability rules, the uncertainty relations, and so forth), and mathematical (the irreducibility of complex numbers, the role of operators, the transformation theory, and so forth). Both Dirac's and von Neumann's schemes are more or less automatic translations of Heisenberg's matrix scheme.

Heisenberg's stroke of genius was finding his matrices, and it was itself a founding theoretical move. That is, this arrangement of the relationships between observable quantities in infinite matrices of complex numbers (never observable as such) is already *theory*, not observation of nature, which does not arrange anything in this way. It is nearly a miracle that Heisenberg proceeded to arranging or, as against the classical view, rearranging the available data into his matrices, infinite square tables of complex-number quantities related to these data. (These matrices must be infinite in this case, for example, in order to derive the uncertainty relations.) He was, famously, not even aware at the time that the corresponding mathematical theories (matrix algebra) already existed at the time, which was realized by Born, his teacher, upon reading the paper. Heisenberg reinvented matrix algebra, and reinvented from physics, as stressed by Alain Connes in the context of noncommutative geometry, for which Heisenberg's discovery provided arguably the main point of departure and on which I shall further comment below (Connes 1994, p. 38). For, the key mathematical features of new quantum mechanics, beginning with the new kinematical elements and the way of their multiplication, leading to the noncommutativity of this multiplication, emerged from experimentally established rules governing the data in question.

This fact repeatedly proved its significance in the history of quantum mechanics, first of all, because it facilitated the ability of the theory to coherently account the experimental situations and procedures in question in quantum mechanics. Thus, as will be seen, Bohr's counterarguments to EPR's argument and to related arguments of Einstein were grounded in a careful discrimination between what belongs to the phenomena in questions (e.g., the uncertainty relations) and what belong to a given theory, and how the phenomena and the theory match in quantum mechanics. Furthermore, quantum phenomena are perhaps the greatest demonstration that confirms that nature's "imagination" far exceeds ours. To risk a strong claim, nothing in the preceding intellectual or even human history, in our conscious thinking and imagination or in our dreams, was able to come up with or prepare us for what these phenomena show us in the double-slit experiment, in the delayed choice experiment, or in the EPR correlations, to name just a few among such phenomena. In Wheeler's words describing quantum phenomena: "What could have dreamed up out of pure imagination more magic—and more fitting—that this?" (Wheeler 1993, *QTM*, p. 189) One may indeed wonder: "More fitting" than what or fitting to what, apart from nature itself or (this perhaps Wheeler's meaning, following Bohr), capable, now in Einstein's words, of being "logically possible without contradiction"? This, however, is perhaps only possible at the cost of making the unthinkable, something we cannot possibly explain or conceive of, part of this logic. Our imagination and thinking can, however, lead us to this logic and accompany it by rigorous physics as a mathematical science of nature, which is of course what happened in Heisenberg's discovery of quantum mechanics.

In epistemological and phenomenological terms, the introduction of new mechanics in Heisenberg's paper may be seen as corresponding to and perhaps arising, at least in part, from an extraordinary form of "vision" of the constitution of the data in question in quantum mechanics. In this *view* (in either sense), one divest the quantum-mechanical data, say, spectral lines or (it is easier to use this archetypal experiment) traces left on a silver screen in the double-slit experiment (in both types of set-up), of the presumed classical-like and hence configurable, especially in geometrical terms, history of their emergence. That is, one divests them of any history that could possible be mapped, mathematically or conceptually, by a classical model.

For example, such traces should not be seen either as points resulting from classically conceived collisions between "particles" and the screen or as resulting from a classical wave propagation. Neither "picture" corresponds to what actually occurs. At this stage, even the radical (without a possibility of reconstituting the nature of the object that left a trace) trace-like character of these marks is suspended, although this character will have to be given to these marks in order to treat them in quantum-theoretical terms. In part by virtue of two possible outcomes of the experiment depending upon a chosen setup, the appearance of these marks cannot ultimately be *explained* in these or any terms, but only predicted by means of the quantum-mechanical formalism, properly corresponding to each setup. One does not of course need quantum mechanics to predict qualitatively either the presence or absence of a wave-like interference pattern in the double-slit experiment, depending on whether both slits are open (and no counters installed allowing one to know through which slit each quantum object considered passes). These facts may and indeed must be seen as experimental facts, which we can expect once we set an experiment properly. Quantum mechanics, however, enables one to predict the specific, measurable character in each such experiment (including in its statistical aspects), depending on the parameters involved, which classical physics fails to do. Accordingly, in order for a theoretical formalization to take place, these marks, while "visible," have to be divested of any form of mathematical and specifically geometrical representation as concerns the processes of their emergence.

Classical physics is defined by the possibility of such representation of the situation, making it available to human intuition, defined by a possibility of geometrical visualization, at least in principle, of the processes it considers. This geometry can of course be given its algebra as well, or, depending where one begins, the classical-mechanical algebra can be given a proper geometry. By contrast, quantum-mechanical algebra suspends the possibility of such geometrical visualization even in principle. The traces forming the quantum-mechanical data must be seen as allowing for no classical physical description as concerns the processes, history, of their emergence.

Heisenberg, thus, first suspends any possible picture of the emergence of these data in accordance with classical physics and its geometrical representation of such processes. Instead, he treats these data as "effects" divorced from any classical configurativity and hence also classical (there may be no other) causality as concerns their emergence. Heisenberg does not philosophically explore the epistemological consequences of the situation, of which he was only vaguely aware at the time. His main concern was to offer a mathematical formalism that would enable theoretical predictions in the situations where all previous attempts had failed. These consequences emerged in subsequent developments, both in Heisenberg's own work and in the work of Bohr and others. The nature and significance, physical or epistemological, of Heisenberg's contribution and, again, vision was, however, decisive, one of the great events in the history of the twentieth-century physics.

It is indeed somewhat of a miracle that Heisenberg proceeded from the disassemblage just considered of the experimental data in question to arranging or, as against the classical view, rearranging this data into his matrices with probabilities of transitions from one state of the system (cum measurement) to another as their elements. Of course, once one studied how Heisenberg arrived at his ideas, as partially sketched here, the situation does not look quite so dramatic. It never does. His invention was quite miraculous nevertheless, and in a way more dramatic and remarkable given the mathematical and physical specificity of the process of his discovery. Reciprocally, the logic and epistemology just outlined also support and explains some of Heisenberg's seemingly unmotivated radical steps on his difficult and protracted, although all things considered, not so long a path to his discovery, without of course diminishing the revolutionary nature of these steps, and without necessarily unduly linearizing this path into a single determined sequences. Many things happened along the way, via many trials and errors, moving back and forth, and so forth.[11]

From this perspective, Heisenberg, rather than only Bohr, may be seen as the discoverer of the radical epistemology of quantum mechanics, albeit without realizing the ultimate limits of this epistemology. Nor of course does one find in his work or related work on matrix quantum mechanics (or on Schrödinger's wave mechanics) complementary features of description or such accompanying concepts, such as individuality, phenomena, effects, and so forth, leading, among other things, to the nonclassical architecture of quantum-mechanical information. All these features came later courtesy of Bohr, some of them much later.

---

[11] See volumes 2 and 3 of Mehra and Rechenberg's *The Historical Development of Quantum Theory* (Mehra and Rechenberg 2001) for an extended historical account of Heisenberg's discovery and the development of matrix mechanics.

As I argue here, complementarity itself, as an interpretation of quantum mechanics, had undergone considerable evolution before it reached, in the late 1940s, its ultimate version, presented in "Discussion with Einstein." Sometime between 1935 and 1938, under the impact of Einstein's criticism, and specifically the EPR 1935 argument, Bohr rethought the very concept of phenomenon as applicable in quantum physics.[12] As can be easily seen, however, especially in the Warsaw lecture of 1938, "The Causality Problem in Atomic Physics," where the concept was introduced, Bohr's post-EPR thinking is accompanied by his return to the terms of Heisenberg's version of the theory (*PWNB* 4, pp. 94-121). Phenomena would no longer refer to physical space-time properties of quantum objects or their behavior, but to experimentally observed *effects* manifest in, in terms of their physics, classically described measuring instruments under the impact of their quantum interactions with quantum objects (*PWNB* 2, p. 64). It would be more accurate to see "phenomena" as referring to the representations of such effects, and Bohr's usage can indeed be adjusted accordingly. The rigorous specification of each experimental arrangement is essential, since it is itself considered part of each phenomenon and is one of the reasons for the complementary (mutually exclusive) character of some of them. Complementary features of description are defined by the mutual exclusivity of some among experimental arrangements, considered, nevertheless, as each equally possible at any given point, and as all necessary for a comprehensive description of all quantum phenomena, or, again, for defining the *possible* information concerning the objects, that is, their effects upon the measuring instruments.

It follows that any information (in either sense) obtainable by engaging with quantum processes (or processing) can only pertain to the phenomenal effects in question, not to quantum objects themselves, which entail a new form of epistemology, as just outlined, and, by implication, a new form of information processing. One can *relate* the information in question to and correlate it with quantum objects, but one cannot assign it to the latter in terms of properties of such objects. This information itself concerning every individual effect is classical *in its physical nature* (and it can be recorded and communicated accordingly, in classical "bits," for example), but it is very special and (in terms of the physics involved) inherently nonclassical *in structure or architecture*, that is, in the organization of such individual records. This circumstance is crucial for quantum cryptography and computing, and indeed (in this interpretation) makes them possible. It also follows that in this interpretation (in contrast to other interpretations,

---

[12] These changes in Bohr's views have been noticed, but rarely adequately considered. Indeed sometimes they have been used to criticize Bohr's argument or "the spirit of Copenhagen" in general, as, for example, by Mara Beller in *Quantum Dialogues*, whose main aim, however, is an advocacy of Bohm's hidden-variables approach (Beller 1999). As I have argued previously, Beller's argument appears to me unconvincing and to be missing some of the essential points of both Bohr's views, earlier or later (Plotnitsky 2002, pp. 254-255, n.33).

such as hidden variables or the many-worlds interpretations) this *possible* information not only can never be actually available but, more radically, cannot be assumed to be, while unavailable, actually physically existing anywhere.

By the same token, the mathematical formalism of quantum mechanics refers, in a particular statistical way, to these effects and only to them, rather than describes or otherwise accounts for the physical processes (of quantum nature) that lead to the emergences of such effects. The quantum-mechanical predictions concern only such effects, and are made on the basis of these effects, and such predictions cannot be made by means of classical physics, but quantum theory itself does not describe or, again, in any way account for the quantum processes themselves.

4. COMPLEMENTARITY, PHENOMENA, AND THE DOUBLE-SLIT EXPERIMENT

I would like to illustrate the preceding argument by discussing from the perspective this argument entails, the double-slit experiment—the "archetypal" quantum-mechanical experiment, which may be argued to contain most of the key features of quantum phenomena and most of the questions these features pose. The well-known arrangement consists of a source; a diaphragm with a slit (A); at a sufficient distance from it a second diaphragm with two slits (B and C), widely separated; and finally, at a sufficient distance from the second diaphragm a screen, say, a silver bromide photographic plate. A sufficient number (say, a million) of particles, such as electrons or photons, emitted from a source, are allowed to pass through both diaphragms and leave their traces on the screen. Provisionally, I speak for the moment in terms of quantum objects themselves. Strictly speaking, we can only observe certain effects on the screen or physically equivalent macro-phenomena. Indeed all we have in any given event is a final trace of a "collision" on the screen. Everything else, the emission of the particle, its passing or not passing through slits, and so forth, is ascertained on the basis of other observations and measurements that we can perform in similar circumstances, each leading to an individual phenomena in Bohr's sense. Two set-ups are considered: in the first we cannot know through which slit each particle passes, in the second we can, at least in principle (a qualification of considerable importance).

If both slits are open and no arrangements, such as particle counters, are made that would allow us to establish through which slit each particle passes, a "wave-like" interference pattern will emerge on the screen. In principle, this pattern will emerge regardless of the distance between slits or the time interval between the emissions of the particles. The traces of the collisions between the particles and the screen will "arrange" themselves in a pattern even when the next emission occurs after the preceding particle is destroyed after colliding with the screen. This pattern is the actual manifestation and, according to Bohr's and most standard interpretations, the only possible manifestation of quantum-mechanical "waves." In this type of interpretation at least, one can speak of "wave-propagation" or of any attributes of the classical-like phenomenon of wave-propagation (either associated with individual particles or with their behavior as a multiplicity) prior to the appearance of these registered marks only by convention. As noted from the outset, however, according to this interpretation, the same is also true as concerns the attributes of classical particle motion, in particular trajectories. We see on the screen only classically manifest traces of the collisions between quantum objects and the screen. The latter themselves are destroyed in the process of this, in Bohr's terms, "irreversible amplification" of all our encounters with quantum objects to the classical level (*PWNB* 2, p. 51; *PWNB* 3, p. 3).

If, however, there are counters or other devices that would allow us to check through which slit particles pass (merely setting up the apparatus in a way that such knowledge would in principle be possible would suffice), the interference pattern inevitably disappears. The fact that even the possibility in principle of knowing through which slit the particles pass would inevitably destroy the interference pattern may be shown to be equivalent to uncertainty relations and may be given a statistical interpretation (*PWNB* 2, p. 43-47; *PWNB* 4, pp. 76-77). Indeed, as a result, the statistical nature of quantum mechanics appears to be irreducible (*PWNB* 2, p. 34). It is itself an aspect of "the peculiar individuality of quantum effects," the concept to be discussed in more detail in Chapter 3. In other words, we do not know and, in Bohr's view, ultimately cannot know why certain probability rules, say, those of Planck's law or Born's square moduli rule for calculating quantum-mechanical probabilities in fact apply, but they do. Whenever the interference pattern is found, one cannot assign probabilities to the two alternative "histories" of a "particle" passing through either one slit or another on its way to the screen. If we do, the quantum-mechanical probability sum law (based in adding amplitudes rather than probabilities) would not be obeyed and the conflict with the interference pattern will inevitably emerge, as Bohr stressed on many occasions, including those just referred to.[13] One can also put it as follows. In counting the probability involved, we must, as it were, take into account the possibility of a particle passing through both slits, when both are open to it, in calculating the probabilities of the outcomes of such experiments, otherwise the probability will not come out right. We cannot, however, again, at least in Bohr's interpretation, assume that a particle can actually do so in space and time. Hence, the probability counting procedure in quantum physics is different from that of classical statistical physics. I shall return to the subject of probability in quantum mechanics in Chapter 4.

---

[13] It is easy to see that, if (at least given locality) we assume that complementary variables, while subject to uncertainty relations, nevertheless pertain to quantum objects, quantum-mechanical probabilities will not come out right, as, for example, Peter Mittelstaedt observes (Mittelstaedt 1987, p. 232). My argument here is that Bohr's interpretation entails an even more radical epistemology, whereby not even single properties can be ascribed to quantum objects themselves.

The situation just described, sometimes also known as the quantum measurement paradox, is remarkable. Other standard locutions include strange, puzzling, mysterious (and sometimes mystical), and incomprehensible. The reason for this reaction is that, if one speaks in terms of particles themselves, in the interference picture the behavior of each particle appears to be "influenced" by the location of the slits. Or, even more radically, individually or (which is hardly less troubling) collectively, particles appear somehow to "know" whether both slits are or are not open, or whether there are or are not counting devices installed. Any attempt to picture or conceive of this behavior in terms of physical attributes of quantum objects themselves appears to lead to a logical contradiction; to be incompatible with one aspect of experimental evidence or the other; or entail strange or mysterious behavior; to require more or less difficult assumptions, such as attributing volition or personification to nature in allowing particles individual or collective "choices," as Bohr points out (*PNWB* 2, pp. 51,73); or to imply nonlocality, as instantaneous connections between spatially separated events, which would be incompatible with (special) relativity. The latter alternative was first proposed by Einstein and is legitimate and, arguably, the most rational under this attribution. Yet another alternative would be a retroaction in time, which is hardly more palatable, at least to this author and to most physicists, but is entertained by some.[14]

Bohr, by contrast, sees the situation as indicating, in his ultimate language, the "essential ambiguity" of ascribing conventional physical attributes to quantum objects themselves or their independent behavior. He writes: "To my mind, there is no other alternative than to admit that, in this field of experience, we are [rather than with properties of quantum objects] dealing with individual phenomena [in Bohr's sense] and that our possibilities of handling the measuring instruments allow us only to make a choice between the different complementary phenomena we want to study," that is, two different types of effects of the interaction between quantum objects and measuring instruments upon those instruments (*PWNB* 2, p. 51). The term phenomena is here best understood in Bohr's sense, indicating two different types of effects of the interaction between quantum objects and measuring instruments upon the latter. As Bohr's clarifies later in "Discussion with Einstein," referring to the Warsaw lecture, "The Causality Problem in Atomic Physics" (1938):

> I advocated the application of the word *phenomenon* exclusively to refer to the observations obtained under specified circumstances, including an account of the whole experimental arrangement. In such terminology, the observational

---

[14] Henry Stapp is one of the advocates of this view, which, he argues, is a consequence of quantum mechanics. See (Stapp 1989) and (Stapp 1997). For an effective counterargument to the second article, via Bohr's reply to EPR, see (Mermin 1998b). That retroaction would allow for a realist interpretation of quantum mechanics is a different type of argument, also found in recent discussions.

> problem is free of any special intricacy since, in actual experiments, all observations are expressed by unambiguous statements referring, for instance, to the registration of the point at which an electron arrives at a photographic plate. Moreover, speaking in such a way is just suited to emphasize that the appropriate physical interpretation of the symbolic quantum-mechanical formalism amounts only to predictions, of determinate or statistical character, pertaining to individual phenomena appearing under conditions [of measuring instruments] defined by classical physical concepts. (*PWNB* 2, p. 64)

The term "phenomenon," thus, now only refers to registered observations or measurements, in other words, only to what has already happened and not to what may happen, even if the latter possibility corresponds to a rigorous prediction enabled by quantum mechanics, which, again, could only be statistical and hence never fully guarantee a given outcome. This fact is in part what leads Bohr to his emphasis on the actual in defining "phenomenon." This centrality of the actual in Bohr's definition of phenomenon is a crucial point, often missed by commentators, specifically in the context of the EPR argument. I would say *an* (rather than *the*) appropriate interpretation. On the other hand, it is important to note that, contrary to yet another common misunderstanding, Bohr's appeal to the necessity of classical physical concepts only involves the description of certain parts of measuring instruments and their arrangements, but not quantum objects themselves or their quantum interactions with other parts of measuring instruments. Quantum objects and processes, including these interactions, are not describable by means of classical physical concepts or by means of any concepts. These interactions are, however, capable of producing physical changes at the classical macro-level in the measuring arrangements involved. They are, as it were, amplified to the classical level, compelling Bohr to speak of "practically irreversible amplification effects" (*PWNB* 2, p. 51).

These circumstances deprive us, on the one hand, of the possibility of either further subdividing or otherwise rearranging phenomena so as to avoid complementarity at that level, and on the other, of the possibility of seeing complementarity as arising (while inevitable in practice) from something that is, in principle, free of complementarity (for example, in the manner of hidden variables). Accordingly, Bohr argues that "the problem again emphasizes the necessity of considering the *whole* experimental arrangement [or, again, the phenomenon], the specification of which is imperative for any well defined application of the quantum-mechanical formalism" (*PWNB* 2, p.57). As he also notes elsewhere:

> Within the scope of classical physics, all characteristic properties of a given object can in principle be ascertained by a single experimental arrangement, although in practice various arrangements are often convenient for the study of

> *different aspects* of the phenomena. In fact, data obtained in such a way simply supplement each other and can be combined into a consistent picture of the behavior of the object under investigation. In quantum physics, however, evidence about atomic objects obtained by different experimental arrangements exhibit a novel kind of complementary relationship. Indeed, it must be recognized that such evidence, which appears contradictory when combination into a single picture is attempted, exhausts all conceivable knowledge about the object. Far from restricting our efforts to put questions to nature in the form of experiments, the notion of *complementarity* simply characterizes the answers we can receive by such inquiry, whenever the interaction between the measuring instruments and the objects forms an integral part of the phenomena. (*PWNB* 3, p. 4; emphasis added)

As I noted in the introduction, this argumentation makes Bohr's use itself of the term "complementarity" peculiar, since the word complementary usually indicates parts of or adding up to a whole, which is by definition impossible under the conditions of complementarity in Bohr's sense. This difference is here captured and explained by Bohr's reference to "different aspects of the (same) phenomena" in classical physics, as opposed to two different, mutually exclusive or complementary, phenomena in quantum physics. The latter phenomena are never reducible to or derivable from a single entity of any kind, which gives a very peculiar character to the efficacity behind the quantum-mechanical effects. But this is also what gives the concept a well-defined meaning and what enables a consistent and rigorous interpretation of quantum mechanics. If certain aspects of the situation seem inconceivable, this is because, in Bohr's view, they are. They certainly are not available to anything conceived on the model of classical physics, but more radically on any model that is or possibly ever will be available.

Once this type of interpretation is in place and reference to the properties of quantum objects themselves is suspended, the undesirable features mentioned above are removed without affecting the integrity of the data or the formalism of quantum theory. This is a reasonably trivial application of the standard logical deduction under nontrivial, and to some unacceptable, epistemological assumptions. As will be seen, a similar argument (concerning counterfactual reasoning) allows one to ascertain the locality of quantum mechanics. At the very least, Bohr's interpretation is both consistent with the quantum-mechanical predictions and local.

In this interpretation, an unambiguous reference to quantum objects or processes themselves would remain impossible even when one speaks of single such attributes, rather than of a simultaneous attribution of joint properties involved in various uncertainty relations, and even at the time when the measurement takes place. This stronger claim or, again, the corresponding interpretation appears to have emerged in Bohr's arguments in the wake of EPR's article. It is clear, however, that this is what

Bohr has in mind, since in speaking of this ambiguity he never qualifies it by a reference to either joint properties or to the uncertainty relations. All quantum measurement is in effect defined by Bohr in terms of correlations that have no specifiable quantum correlata (at either end, that of the object and that of the quantum constitution of the instrument), but only *one* specifiable classical correlatum, pertaining to the classically describable aspects of the instrument. As Bohr says:

> This point is of great logical consequence, since it is only the circumstance that we are presented with a choice of *either* tracing the path of a particle *or* observing interference effects, which allows us to escape from the paradoxical necessity of concluding that the behavior of an electron or a photon should depend on the presence of a slit in a diaphragm through which it could be proved not to pass. We have here to do with a typical example of how the complementary phenomena appear under mutually exclusive experimental arrangements ... and are just faced with the impossibility, in the analysis of quantum *effects*, of drawing any sharp separation between an independent behavior of atomic objects and their interaction with the measuring instruments which serve to define the conditions under which the phenomena occur. (*PWNB* 2, pp. 46-47; emphasis added)

We now see why Bohr needs his concept of phenomena as defined by the appearance of the particular individual effects recorded in certain parts of measuring instruments under rigorously specifiable experimental conditions and why this specification must itself be seen as part of the phenomena (*PWNB* 2, p. 64). In Bohr's interpretation, the constitutive role of these conditions can be never eliminated in considering the outcomes of quantum-mechanical experiments (in the way it can, at least in principle, be done in classical physics), even when, as in the EPR experiment, our predictions concern objects that are physically unaffected by measurements enabling such predictions. If seen independently of the quantum-mechanical context of its appearance, each mark on the screen in the double-slit experiment would be perceived in the same way or as the same phenomena in the sense of the philosophical (say, Husserl's) phenomenology. Such a mark would appear the same regardless of the difference in the conditions and, hence, outcome ("interference" or "no interference") of the double-slit experiment. According to Bohr's understanding, however, each mark is a *different individual* phenomenon depending on these conditions, which are always mutually exclusive in the case of complementary phenomena, and are unique in any circumstances. (Quantum-mechanical predictions, including numerical statistical predictions, crucially depend on this distinction as well.)

Thus, in the double-slit experiment, rather than dealing with two phenomena, each defined by a different multiplicity of spots on the screen, we deal with two distinct multiplicities of individual phenomena, defined by each spot. Each of the latter is indivisible—two sets of phenomena in Bohr's sense—depending on two different sets of conditions of the experiment. One of these sets of conditions will lead to the emergence of the interference pattern, "built up by the accumulation of a large number of *individual* processes, each giving rise to a small spot on the photographic plate, and the distribution of these spots follows a simple law derivable from the wave analysis" (*PWNB* 2, pp. 45-46). Each single spot must, however, be seen as a different type of phenomenon, defined by the conditions in which the event occurs. Two different overall patterns, "interference" and "no-interference," pertain to two (very large) sets of different types of individual phenomena.

Far from being a matter of convenience, this distinction between two multiple-spot phenomena and two multiplicities of different individual spot-like phenomena is essential for Bohr's meaning and the consistency of his argumentation. First, no paradoxical properties, such as simultaneous possession of contradictory wave-like and particle-like attributes on the part of quantum objects themselves, are involved. Secondly, and most crucially, in our analysis we can never mix considerations that belong to complementary experimental set-ups in analyzing a given experimental outcome, even when dealing with a single spot on the screen, as we could, in principle, do in the case of classical physics. This is not an uncommon error, including in some of Einstein's arguments, which could indeed lead to the appearance of paradoxes. These paradoxes, however, disappear once this rule of complementary mutual exclusivity of such considerations is followed. Throughout his arguments with Einstein, Bohr stresses that in such situations, which are invoked in most of Einstein's arguments, including of the EPR type, "we must realize that … we are not dealing with a *single* specified experimental arrangement, but are referring to *two* different, mutually exclusive arrangements" (*PWNB* 2, p. 57).

As will be seen in Chapter 3, the point becomes especially significant in the EPR context, in which the term phenomenon in Bohr's sense could only refer to a single measurement actually performed on one and only one of the two particles of the EPR pair, rather than, as some argue, to any physical considerations involving both particles. In the latter case, nonlocality could emerge even in Bohr's scheme, as some argue by, in my view, displacing Bohr's concept. By contrast, the individuality of Bohr's phenomena allows one to avoid nonlocality.

5. FROM BOHR'S ATOMS TO QUBITS

It was, then, Heisenberg's "kinematics" and its implications, most particularly the uncertainty relations, that eventually led Bohr, via complementarity and the argument with Einstein, to a new radical conception of atomicity, which may be given the name

"Bohr's atom" more fittingly than his 1913 conception of the hydrogen atom, the starting point on this long road. I shall now outline this conception, although one may succinctly state it as follows: Bohr's individual phenomena are Bohr's "atoms."

On the one hand, on Bohr's view, quantum objects are irreducibly inaccessible to us, are beyond any reach (including, again, as objects), and in this sense there is irreducible rupture, *discontinuity*, arguably the only quantum discontinuity in Bohr's epistemology. On the other hand, they are irreducibly inseparable, *indivisible* from their interaction with measuring instruments and the effects this interaction produces. This situation may seem paradoxical. It is not, however, if one accepts Bohr's epistemology, according to which the processes responsible for quantum effects are both reciprocal and, hence, indissociable from their effects and yet are outside any knowledge or conception, continuity and discontinuity among them (*PWNB* 2, pp. 39, 62). Bohr's concept of the indivisibility or wholeness of phenomena allows him to avoid the contradiction between indivisibility and discontinuity, and to establish atomicity at the level of phenomena (*PWNB* 2, pp. 40, 72-73; *PWNB* 3, p. 4).[15]

These individual phenomena become Bohr's interpretation of the quantum "atomicity" (in the original Greek sense of being indivisible any further), rather than indivisible atomic quantum objects, as would appear in the wake of Planck's discovery (and as it is still often seen). Quantum objects themselves are no longer assigned and cannot be assigned any features of atomicity anymore than any other features, properties, images, and so forth. Since it is impossible to consider the quantum objects independently of this interaction, this "indivisibility" makes it impossible to isolate quantum objects from their phenomenal enclosure. Bohr sometimes speaks of "closed phenomena" in this sense (*PWNB* 2, p. 71). Any attempt to "open" or "cut through" a phenomenon can only produce yet another closed individual phenomenon, a different "Bohr's atom" or set of such "atoms," leaving quantum objects themselves again (and correlatively) irreducibly inaccessible inside phenomena. As Bohr writes, "the individuality of the typical quantum effects finds its proper expression in the circumstances that any attempt of subdividing the phenomena will demand a change in the experimental arrangements introducing new possibilities of interaction between objects and measuring instruments," which "reveal the ambiguity in ascribing customary physical attributes to atomic objects" (*PWNB* 2, pp. 39-40, 51).

By the same token, phenomena become *individual*, each of them—every (knowable) effect conjoined with every (unknowable) process leading to it—unique and unrepeatable. Some of them can be clustered insofar as they refer to the "same" quantum

---

[15] Contrary to some recent claims (e.g., Beller 1999), this concept has nothing to do with David Bohm's concept of "wholeness" as the nonlocality of the quantum world, indebted as the latter may be to Bohr's ideas. Bohr's concept of the wholeness or indivisibility of phenomena is coextensive with the locality of his interpretation, which I shall discuss in Chapter 3.

entities, "individual," such as elementary particles, or collective, for example, such more or less stabilized composites of quarks and gluons, such as protons and neutrons. Reciprocally, however, this offers us the way (and, on Bohr's view, the only way) to define and identify such entities, individually or collectively. Thus, along with the quantum atomicity as *indivisibility*, the quantum atomicity as *individuality* (originally applied to quantum objects) is now also understood as the individuality (uniqueness) of phenomena. This individuality also implies a form of discontinuity or, more accurately, *discreteness*, yet another feature of atomicity, corresponding to the no-continuum postulate, introduced earlier, insofar as each such phenomenon or effect is singular and indeed unique, and as such is separated or isolated from any other such phenomenon or effect. Bohr's quantum *discontinuity*, while equally different from discontinuity at the level of quantum objects, is more properly seen as the impossibility of applying either of these concepts, continuity or discontinuity, or any conceivable concept, to their "relation" (another inapplicable concept) to the manifest effects of their (quantum) interaction with measuring instruments, which is responsible for these effects.

Thus, all of the standard features of quantum atomicity—individuality, wholeness/indivisibility, discreteness, and discontinuity—are transferred from the level of atomic objects to the level of instrumental, technological phenomena in Bohr's sense. This argument explains why Bohr argues that quantum physics reflects a new feature of atomicity, foreign to the mechanical (atomic) conception of nature, extending from Democritus's atomism. The argument is invoked as early as 1932 (*PWNB* 2, p. 6), but becomes especially prominent and persistent in Bohr's later writings. The point is stated twice at the outset of "Discussion with Einstein": "Planck's discovery … revealed a feature of atomicity in the laws of nature going far beyond the old doctrine of the limited divisibility of matter … Einstein … explored with a most daring spirit the novel features of atomicity which pointed beyond the framework of classical physics" (*PWNB* 2, p. 33; also *PWNB* 2, pp. 25, 71; *PWNB* 3, p. 2; *PWNB* 4, p. 94). Classical physics is based on this traditional doctrine, as the latter is in turn reciprocally based on classical physics or proto-physics, such as that of Democritus or Epicurus and Lucretius. This physics would describe the behavior of its objects as such in, at bottom, a causal (even if practically statistical or entirely unpredictable, for example, chaos-theoretical) way. In Bohr's interpretation, the only ultimate "atomic" entities that can be rigorously described are certain indivisible configurations of experimental technology.

This is one of Bohr's greatest and most extraordinary conceptions to which the term "Bohr's atom" may apply best. As I noted, these "atoms" are not physically atomic, since they are composed of more elementary (quantum) physical constituents, but they are atomic in the sense that no different form of access to quantum objects is possible, regardless of how far we can divide and subdivide our experimental technology, or matter itself. Conversely, however, all "elementary" quantum objects could only be defined through these indivisible macro-atoms. These circumstances prevent any possibility for quantum objects to appear or to be considered, or even conceived of, in any specific, such as "atomic," form independently, outside of the instrumental, technological *enclosures* of specific experiments, of "[en]closed phenomena" (*PWNB* 3, p. 73).[16]

Bohr's technological conception of atomicity also gives a proper meaning to Bohr's argument concerning the irreducible discrimination "between the *objects* under investigation and the *measuring instruments* which serve to define, in classical terms, the conditions under which the phenomena appear" (*PWNB* 2, p. 50). Bohr sees "this necessity of discriminating in each experimental arrangement between [them]" as "a *principal distinction between classical and quantum-mechanical description of physical phenomena*" (*PWNB* 4, p. 81). For, while "in classical physics the distinction between object and measuring agencies does not entail any difference in the character of the description of the phenomena concerned," in quantum physics it does (*PWNB* 4, p. 81; also *PWNB* 2, p. 50; *PWNB* 3, p. 3).

This and related statements by Bohr may suggest that while parts of measuring instruments are described by means of classical physics, the behavior (in space and time) of quantum objects is described by means of quantum-mechanical formalism. Bohr obviously says the former, but he clearly does not ever say and does not mean the latter. It is an immediate consequence of Bohr's argumentation that the behavior of quantum objects cannot be seen in terms of any formalism, classical or quantum-mechanical. The quantum-mechanical formalism does not in any way describe the ("undisturbed") quantum objects and processes, or their properties, even single properties. It does not do so either before the measurement interference takes place, or between instances of such interference, or even during measurement. Throughout his writing, from the Como lecture on, Bohr rejects the view that the quantum-mechanical formalism can unambiguously refer to quantum objects and processes in terms of space-time concepts independently of measurements, and in his post-EPR works even at the time of measurement. As he said, "even in the indeterminacy relation[s] we are dealing with an implication of the formalism which defies unambiguous expression in words suited to describe classical physical pictures" (*PWNB* 2, p. 40). Ultimately this formalism defies all "unambiguous use of space-time concepts," which are "confined to the recording of observations which refer to marks on a photographic plate or similar practically irreversible amplification effects" (*PWNB* 2, p. 51). Bohr saw the mathematical formalism of quantum mechanics as correlative to this situation.

Already in the wake of Heisenberg's introduction of matrix mechanics, Bohr saw the latter as establishing new relationships between mathematics and physics, and

---

[16] Bohr's conception of atomicity and its significance have rarely been properly examined. One exception is Henry Folse (e.g., Folse 1987). Folse's interpretation of this conception and of Bohr's epistemology is, at least on this occasion, different from the one proposed here. In particular, he appears to attribute to Bohr the suspension of locality, which argument is, as I said, difficult to sustain. Folse's more recent arguments appear to depart from this view (Folse 2002).

(not a view customarily associated with Bohr) stressed the significance of mathematics in "prepar[ing] the way for further progress" (*PWNB* 1, p. 51). So, of course, even more emphatically, did Heisenberg himself throughout his life. Thus, the difference between the two descriptions is that classical physics describes the classical world, including certain parts of the measuring instruments, while quantum physics relates to the impact of the interaction between quantum objects, indescribable by means of either classical or quantum theory, upon certain (classical described) parts of measuring instruments.

Reciprocally, Bohr's point bears crucially on the question of the classical description of the measuring instruments involved. For, as stressed earlier, rather than to the constitution of measuring instruments as a whole (as is often assumed by Bohr's interpreters), the classical description applies only to the effect-constituting parts of these instruments. The constitution of the instruments, however, also contains a quantum stratum, through which the interaction with quantum objects takes place. It is this interaction that is responsible for the classical effects in question, that is, for the classical behavior of those parts of measuring instruments that, in their quantum interactions with quantum objects, produce these effects. This quantum stratum and these interactions are subject to quantum-mechanical treatment, including uncertainty relations, and Bohr's epistemology. In short, the measuring apparatus in Bohr is not simply classical, as is commonly argued, but both classical and quantum.

The overall situation has far-reaching implications for both quantum epistemology and the specific character of the mathematical formalism of quantum mechanics (in particular, its dependence on complex rather than real mathematical entities). The quantum-mechanical formalism is highly symbolic and nonvisualizable in its nature, especially by virtue of using complex numbers and infinite-dimensional mathematical magnitudes. This formalism could be related to observations, always recorded in real numbers, by means of artificial schemes, such as Born's square moduli rule for deriving probabilities from quantum amplitudes, von Neumann's projection postulate, and so forth. Accordingly, in Bohr's interpretation, one could speak of so-called linear superposition of "states" in referring to quantum processes themselves only by convention or symbolically, without assuming an actual description of such processes in space and time.

Consider a typical (and typically problematic) representation of the opposite view by Roger Penrose: "A particle's state can involve superposition of two or more different locations. ... a photon's state is such that it can be located in two different beams simultaneously after it encounters a half-silvered mirror" (Penrose 1994, pp. 277-78). On Bohr's view, this statement would be contradictory or, at least, inaccurate. There is no physical evidence that such an "event" could physically occur or be possible, anymore than it is possible for a single photon to pass through both slits in the double-slit experiment. Besides, there are no "beams" in the case of a single photon. Penrose also speaks, similarly inaccurately, of the state of (quantum) spin as "being about several axes at once" (Penrose 1994, p. 278).

There is nothing wrong with the formula itself, |B> + *i*|C>, in question Penrose's statement, in which the ket-vector (a mathematical object in a Hilbert space) |B> corresponds (from the present perspective, symbolically rather than in terms of physical description) to the case of the photon passing through the mirror, while |C> corresponds to that of the photon reflected by the mirror. However, from the present, Bohrian, perspective this can only mean that the photon, that is, the trace of the "photon," could be detected in either "state" with an equal probability, calculated in the standard way from the formula in question, rather than, as Penrose suggests, that the photon can ever assumed to be in both "states" at once, either prior to the detection procedure or if the detection is not performed. In the latter case we simply do not know where it could in principle be found, and there is no possible information prior to such a detection that would allow us to know where it could possibly be. This, however, is not the same as to say that it could ever be simultaneously in both states or places. The same argument applies to spin, in which case the quantum-mechanical formalism of the same (ket-vector) type gives the probability of detecting "spin" in a given direction. Or, again, more accurately one should speak of the effect of spin, without ever being able to ascertain this direction prior to an experiment, but also without presupposing that a quantum object spins about several axes at once. The corresponding mutual exclusivity of measuring procedures and measured effects is epistemologically the same as that of the standard position and momentum measurement, involved in uncertainty relations.

From this viewpoint, the so-called "qubit" (quantum bit), often configured in terms of spin, would indeed refer to the simplest non-trivial 2-level quantum system of the type just described and prepared so that the outcomes of possible experiments can be predicted by using (in addition to the preparation information) the ket-vector formalism as $|\psi> = A\ |0> + B|1>$. (|0> and |1> are chosen basis vectors in a Hilbert space, and A and B, are complex numbers.) While the *possible* information concerning even the simplest quantum system may correspond to arbitrarily large amounts (e.g., numbers of bits) of classical information, the totality of this information is subject to a complementary architecture. This architecture tells us (via Holevo's theorem, which may indeed be seen as a consequence of this architecture) that the maximal amount of information that can be actually obtained or, on the present view, more accurately, generated from a qubit through measurement is one classical bit (Holevo 1973). Accordingly, on the one hand, the possible amount of information that a quantum system, even the simplest one, can generate exceeds immeasurably the capacities of comparable (in terms of their preparation) or indeed any finite classical system. On the other hand, the actual information generated in each case is severely limited by measurement. However, from the present perspective, one would not be able to speak, as it is done sometimes, of the quantum information as contained in a quantum state, such as that of a qubit, but hidden and inaccessible to extraction beyond a single bit. In other words, there is no quantum information, anymore than there are (physical) quantum states, described by quantum formalism or, as Bohr often said, anything quantum, that is anything that could be described as "quantum" in any given sense at the level at which we locate the objects and processes which we refer to as "quantum." There is only classical information generated through the quantum interaction between quantum objects and measuring instruments, and this information has a particular architecture of informational possibilities, defined by their "ordered" or correlational character.

The epistemology of quantum-mechanical information advocated here would not, however, inhibit the possibilities that quantum-mechanical information processing can offer and the advantages it can have over classical information processing, for example, in quantum cryptography and computing. Indeed this epistemology appears to be strictly correlative to the predictive and informational, rather than descriptive (with respect to the behavior of the ultimate objects of the theory, quantum objects), character of quantum physics. How is this information made possible by nature? Bohr's answer is that we may not know how this ultimately is or can be (physically) possible, indeed we cannot know, or even conceive of it, since any further analysis that would, in principle, allow us to do so is "*in principle*, excluded."

6. BOHR'S EPISTEMOLOGY AND DECOHERENCE

One might see an anticipation of the idea of decoherence in Bohr's concept of "irreversible amplification," correlative to and part of the conceptuality of the phenomena and individual quantum effects, as here considered. For, as we recall, these effects are "irreversible implications effects," ineluctably left in a system and, thus, allowing us come to an agreement concerning the data. In this sense, quantum mechanics is objective and observer (but not an observational set-up) independent (*PWNB* 2, pp. 73-74, 90-91; *PWNB* 3, pp. 6-7). As Bohr says, "it must be realized that … all unambiguous use of space-time concepts in the description of atomic phenomena is confined to the recording of observations which refer to marks on photographic plates or to similar practically *irreversible amplification effects* like the building of a water drop around an ion in a cloud chamber" (*PWNB* 2, p. 51; emphasis added). What is now called "decoherence" is concerned with the physics of this process and more generally with the transition from the quantum level to the macro-level of the world that we experience. Bohr's appeals to "irreversibility" also indicate the potential thermodynamic aspects of decoherence, which have been part of the theoretical discussions concerning decoherence (e.g., Zurek 2003). Most decoherence theories exhibit marked affinities with Bohr's interpretation. Thus, the macro level, the level of phenomena in Bohr, is by and large described in terms of classical physics, keeping in mind the quantum aspects of the constitution of measuring instruments and their quantum interaction with quantum objects, as considered earlier.[17] On the other hand, the processes occurring at the quantum level are presumed by most decoherence theories to be described by the formalism of quantum theory, the view adopted, for example, in Zurek's work just cited. By contrast, as I argue here, in Bohr's interpretation the quantum-mechanical formalism could not be seen as unambiguously referring to the properties of quantum objects and processes in terms of spatial and temporal concepts in the way it can be done in classical physics or even in relativity. Indeed, as Bohr notes throughout his writings, relativity already introduces significant complications in this respect.

Accordingly, in Bohr's view, while something like "decoherence" ("irreversible amplification") would take place, the formalism of quantum theory would refer , in terms of predictions, to certain classically manifest "effects," but, it follows, not the physical processes responsible for these effects. One might say, if one wants to apply such terms to Bohr's argumentation, that for Bohr there is decoherence, but there is no (physical) "coherence" that decoheres. That is, one could speak of linearly superimposed ("coherent") states in referring to quantum processes themselves (that of a single "particle" or of an extended quantum system) only by convention or symbolically, without assuming an actual "coherent" (or any other) description of such processes in space and time. Such "states" will, of course, manifest themselves in particular phenomenal effects. In this view, there could be no *physical "coherence"* or anything that can be conceived of in terms of space-time processes corresponding to the coherence (say, again, a linear superposition of vectors in a Hilbert space) of the mathematical quantum-mechanical description, which would then "decohere" in the process of measurement.[18] The mathematical "coherence" of our description itself "arises" or applies due to the possibility and effects of measurement or preparation, which involves the same type of irreducible and ultimately uncontrollable interaction between quantum objects and measuring instruments, and in this structural sense is a measurement. (In this interpretation, quantum mechanics is, again, a theory that allows one to predict the outcomes of future possible measurements on the basis of measurements already performed, and only such outcomes, rather than to say anything about the state of the system between measurements.) In Bohr's view, this interaction both allows for the

[17] In addition, as I noted in the introduction, there are also macroscopic quantum effects, meaning this now not so much in Bohr's sense just indicated (although this sense would apply and, as will be seen presently, is especially appropriate), but in terms of their size as macroscopic "coherent" quantum systems. Their behavior, however, is defined by their microscopic quantum constitution.

[18] A given state in a linear superposition is itself subject to this impossibility, as Dirac noted (Dirac 1985, pp. 10-14). For the same reason, the terminology of either the state vector (which concept has no physical meaning in Bohr's framework) or the collapse of the wave function is not favored by Bohr. I shall return to the question of quantum states in the next chapter.

particular mathematization of the overall situation in terms of quantum-mechanical formalism and epistemologically limits the meaning of the latter, which makes this situation correlative to the inevitability of the uncertainty relations in quantum physics. Just as, and, again, correlatively to, the application of specific probabilities to the quantum-mechanical data, the application of this mathematics cannot be explained in the way it would be or could be in classical physics. This irreducible inexplicability does not eliminate "decoherence," if we refer by the latter to the process of transition from the quantum world to the classical world, but it has an impact upon what kind of decoherence theory or its physical meaning one can speak of under these conditions.

Bohr would not, I think, argue that there is nothing more to say concerning the physics of transition from the quantum to the classical level, say, in the building of a water drop around an ion in a cloud chamber, or that this problem as resolved. This may be said even leaving aside Bohr's preoccupation with the correspondence principle, which was, we recall, indispensable to Heisenberg's invention of quantum mechanics and which clearly bears on this problem or rather this problem bears on it. Bohr was well aware that Heisenberg's paper on the uncertainty relations addressed this situation. He was also well aware of the Ehrenfest theorem, the significance of which was so instructively commented on by Omnés in his recent work (Omnés 1994). In Bohr's view, the "gap" between the physical description of macroscopic physical phenomena and quantum objects and processes is insurmountable, whether the macro phenomena in question are, in general, those considered in classical physics or those in question in and described by quantum physics. The latter is, again, seen as describing only the effects of the interaction between quantum objects (which may be macroscopic in terms of the range of their effects) and measuring instruments.

Accordingly, while this view does not suspend the possibility or necessity of such theories, it entails that the "gap" in question would inevitably appear in any proper "decoherence" theory, a theory accounting for this transition. It would also appear in any theory accounting for the ultimate constitution of the macro-world (available to a classical description) out of the ultimate (quantum) micro-constituents of matter. The point may be put as follows. It is possible to develop "macro" theories, say, accounting for the building emergence of a water drop around an ion in a cloud chamber, which is an interesting problem. Such theories may well be of a mixed, quantum and classical nature, including thermodynamics and chaos theory, which, while different epistemologically, are both classical. There could be, however, no truly "micro" account, which would or could account for this "gap," since such an account would entail an ultimate subdivision (and hence de facto elimination) of Bohr's phenomena, which is not possible. Each phenomenal effect in question, such as a trace that appears due to the building of a water drop around an ion in a cloud chamber, is both an actual physical phenomenon and part of the model of the quantum-mechanical situation. This model itself is established once the totality of all such effects is considered and once an access to the underlying dynamics responsible for these effects is "*in principle* excluded." That is, nothing qualitatively different can be observed or be properly related to the mathematical formalism of quantum theory.

This is in part why Bohr argues that the actual impossibility of performing some of the experiments he describes in his arguments is essentially irrelevant. These "experiments" offer us a structural "model," realized in actual observations, such as those related to the Compton effect, one of his favorite examples (*PWNB* 4, p. 76, n.). He clearly saw the Heisenberg-microscope thought-experiment as exemplifying precisely this point—the impossibility of reducing the radical discontinuity in question between "quantum objects" and "processes" (even if these terms, again, apply) and observable macro (phenomenal) effects in any experiment or by means of any description.

As discussed earlier, if there is "quantum discontinuity," this would be the one. Accordingly, similarly to other quantum-theoretical *phenomena* (Bohr's sense becomes especially useful, if not obligatory, here), new experimental and theoretical findings in that domain might be best seen in terms of new quantum effects manifest in measuring instruments. We can observe and can account for them in terms of particular quantum interactions of specific systems (and their quantum subsystems), even though, consistently with the scheme just outlined, we can, in all rigor, never describe the ultimate efficacity of these "effects." Quantum measurement is obviously itself such an effect, as, it follows, is any conceivable observation of "quantum objects," that is, of an effect with which we would associate such objects and from which we infer their existence, to begin with.

In this area of investigation, too, we may be able to reach and in fact have already reached quite far, experimentally and theoretically, and sometimes this epistemology helps us, as Zurek's work especially suggests. Recent, often spectacular, experimental observations of quantum "coherence" would, I would argue, confirm this point, as they may be seen as presenting us with new effects of both "coherence" and decoherence (in this sense), and it appears to me that related claims concerning observations of "coherent" or linearly superposed states are, in all rigor, subject to this scheme (Haroche et al. 1997; Myatt et al. 2000). Accordingly, certain claims, sometimes found in literature, concerning the existence of physically coherent behavior at the quantum level do not seem to be justified, nor are they expressed by the experimenters just mentioned.

Could "decoherence theories themselves be viable, effective, and significant, including for applications, say, in quantum computing? It would be difficult to doubt this, and I do not think that Bohr would. Such theories may, for example, be especially helpful in thinking how quantum computational devices can in principle be realized and actually built in practice. Indeed, similarly to certain interpretations of quantum mechanics, to be discussed below, some of the versions of decoherence may be argued to be consistent with or entail the view just outlined, even though their authors may not quite see it in this way.

In short, in Bohr's interpretation the phenomena or effects (in Bohr's sense) of decoherence would be subject to essentially the same considerations as the *phenomena* of quantum entanglement. The latter may of course itself be seen in terms of scale-extended coherence, while the observed effect may be seen (with the considerations just given in mind) in terms of "decoherence" or irreversible amplification effects, and both effects are often closely linked theoretically and experimentally (Haroche et al. 1997; Myatt et al. 2000). That is, either phenomenon must be seen as phenomena in Bohr's sense, rather referring to the level of quantum objects themselves, while everything just said concerning understanding and investigation, experimental or theoretical, would be retained. It also follows that Bohr's epistemology is in no way incompatible, quite the contrary, with the concept and theories or, more accurately (there are exceptions), most concepts and theories, of quantum computing, including those involving and even based on quantum entanglement. Perhaps it may ultimately help us to put it in place, for example by warning us to stay away from that which is incompatible with the laws of physics, such as and perhaps in particular nonlocality, which is my subject in Chapter 3, but far from exclusively so.

7. THE EPISTEMOLOGICAL LESSON OF QUANTUM MECHANICS

It follows from the preceding discussion that Bohr's interpretation entails a new form of epistemology, applicable well beyond the limits of quantum theory. Hence, Bohr's persistent references to "the epistemological lesson" of quantum mechanics or atomic physics in general (e.g. *PWNB* 3, pp. 74, 92). This epistemology is that of knowledge and conceptualization (only) in terms of effects, while it leaves the nature and character of the ultimate objects and processes responsible for these effects beyond all available knowledge or even conceptualization, and yet understood as essentially, fundamentally responsible for these effects.[19] We cannot assign to these objects and processes any conceivable general structure, say, by analogy with classical physics or in the manner of hidden variables theories, while assuming that some or even all specific aspects of these processes are partially or even altogether unknown. As I noted from the outset of this chapter, in his later works, interactively with "phenomena," Bohr speaks of

---

[19] The apparent proximities and (perhaps, less apparent) differences between this conception and Kant's phenomenal/noumenal conceptuality would require a separate analysis. Certainly, a number of key features of quantum mechanics and Bohr's interpretation, such as and in particular complementarity, give it further specificity not found in Kant, which, it may be argued, also lead to ultimately more radical epistemology as well. On more general relationships between Kant and quantum theory see (Mittelstaedt 1994). I shall comment on this further in Chapter 6.

the key features of quantum data in terms of quantum *effects* throughout, in particular, again, of "the peculiar individuality of the quantum effects," the concept that I shall further discuss in Chapter 4 (*PWNB* 2, pp. 40, 62). Reciprocally, the existence of such irreducibly unknowable or even inconceivable objects and processes can only be inferred from the peculiar character of its effects, for example, the individual character of all effects involved and the complementary character of some of them. A few specific consequences of this view are worth reiterating and further clarifying in closing this chapter. I shall offer a more comprehensive treatment in Chapter 6.

First, while such objects and processes may manifest themselves, in mutually exclusive circumstances, through certain complementary effects, one cannot think a single underlying hidden "wholeness," from which such complementary effects arise. Indeed, one cannot think of the efficacious dynamics in question either single entity governing such complementary effects, either each pair or in their overall totality, nor as multiple in the sense that it could allow one to assign an unambiguous specifically definable, rather than inconceivable, efficacious dynamics to each individual effect. In other words, in this view, such effects are no more subject to a hidden law (e.g., of the type found in hidden variable theories) than it is subject to a manifest law. These efficacious dynamics may be seen as multiple and indeed as irreducibly multiple only in the sense that the (inconceivable) efficacious processes that give rise to each individual phenomenon are each time different. While each time inconceivable, the efficacious dynamic responsible is nevertheless different each time and, hence, is as unique as, and jointly with, the effects produced. This individuality of the (unknowable and inconceivable) efficacious dynamics of each individual effect is a structural, defining part of the *individuality* of phenomena in Bohr, sealed with its envelope as a (forever) closed phenomena (*PWNB* 2, p. 73). We can only read the addresses on such envelopes, as it were, while the ultimate content of letters that nature sends to us are never available to us.

Thus, quantum objects and processes cannot be seen in terms of independent properties, relations, or laws, which, while unavailable, would specifically *define* a certain material entity responsible for giving rise to the (available) effects in question. Instead, it must be seen as reciprocal with and, again, indivisible from its effects: it can never be, in practice and in principle, conceived as isolated, separate from them. Nor, however, can it be seen as fully continuous with these effects. This is not contradictory, given the nature of the epistemology in question, in part defined by the necessity of logically maintaining the features of physical description that would be contradictory otherwise. Instead, all quantum-mechanical and specifically complementary effects appear only within the manifest strata of such indivisible configurations, that is, phenomena in Bohr's sense. The latter, however, also contain the inaccessible and even inconceivable stratum or (we may not be certain) strata. It is this stratum or strata that define the level of quantum objects and processes, including those involving the interaction between quantum objects and measuring instruments, from which the phenomenal effects in question are accordingly discontinuously detached, once the experiments are performed. Quantum objects themselves, as indefinable and inconceivable entities, can be seen as having an independent existence, and must be, for example, in order to ensure the locality of quantum phenomena and quantum mechanics in the EPR situation. They cannot, however, be meaningfully considered apart from their interactions with the measuring instruments. It is through these interactions that they enter the efficacious dynamics in question. Accordingly, in each case, this dynamics is both irreducibly inaccessible and yet irreducibly indissociable from that part of the overall configuration, which is accessible, that is, is subject to Bohr's phenomena and hence is open to representation, conception, knowledge, and so forth. As such, this dynamics is, again, part of the *indivisibility* or *wholeness* of phenomena in Bohr's sense.

This conception is, definitionally, non-realist; it is of course also non-causal, again, definitionally, or automatically, as a consequence of this lack of realism. As I stress here, at best we can say that quantum objects and processes are "real" in the sense that they have actual physical existence, specifically manifest in their capacity to produce certain effects, such complementary phenomena, particular statistical and correlational data, entanglement, and so forth. This capacity, however, allows for no concept of reality that has so far been or will ever be available to us (in particular those borrowed from classical physics) to be applied to it consistently with the effects in question in quantum physics (at least while maintaining the locality of quantum theory). In other words, something in nature gives rise to the effects in question (including when we are not there to observe them interact), but that something is inconceivable in classical and perhaps any possible terms. "Existence" (or "nonexistence") are, too, among such terms, as are the possibility or impossibility of "conceiving" of it, or "possibility" or "impossibility," or "something," or "it," or "is." This may be the ultimate meaning of Bohr's statement that "we are not dealing with an arbitrary renunciation of a more detailed analysis of atomic phenomena but with a recognition that such an analysis is *in principle* excluded," the ultimate epistemological lesson of quantum mechanics.

From this perspective, quantum objects and processes may be even more radically inaccessible than the interior of black holes. While it must be applied with caution, the analogy itself is not out of place, insofar as the "naked" singularity, inevitably (i.e., by virtue of the equations that describe a black hole) found in the interior of a black hole, can never be seen. Black holes must, of course, in turn be seen as quantum objects in view of Stephen Hawking's famous theorems. On the other hand, in certain recent versions of string theory the ultimate constituents of matter are in fact configured as microscopic black holes. This idea is in turn partly due to Hawking, who, actually, suggests that this type of objects may introduce more radical levels of, at least, indeterminacy (but, it appears, also of noncausality and arealism) into physics (Hawking and Penrose 1996).

This limit upon how far our knowledge can reach as concerns the interior of black holes in no way limits our advancement of knowledge concerning them. Nor, accordingly, should we be alarmed by the presence of similar limits in quantum theory. As Bohr stresses, "such argumentation does of course not imply that in atomic [quantum] physics, we have no more to learn as regards experimental evidence [defining these effects] and the mathematical tools appropriate for its comprehension" (*PWNB* 3, p. 6). Nor, one might add, would it imply that quantum-mechanical effects might not be used in practice, for example, in quantum computing. Hence, I only appeal here to the *ultimate* impossibility of knowledge or conception concerning quantum objects and processes. "Indeed," Bohr adds, "it seems likely that the introduction of still further abstraction into formalism will be required to account for the novel features revealed by the exploration of atomic processes of very high energy," that is, in quantum field theory (*PWNB* 3, p. 6). I shall return to the latter subject in Chapter 5 of this study. My point here is that the history of quantum physics has demonstrated and continues to demonstrate Bohr's assessment, made in 1958, as all of its experimental and theoretical findings so far appear to be consistent with Bohr's epistemology.

Chapter 2. Complementarity, Quantum Variables, and the Relationships between Mathematics and Physics

> In quantum theory, a particle also has a precise position and a precise momentum.
>
> –Asher Peres

1. TRANSLATIONS: FROM CLASSICAL TO QUANTUM MECHANICS

As part of his critique of Karl Popper's argument against the Copenhagen interpretation of quantum mechanics in his "Karl Popper and the Copenhagen Interpretation" (Peres 2002), Asher Peres offers a lucid and elegant exposition both of quantum mechanics and of a certain interpretation of it. This interpretation is a *version* of the Copenhagen interpretation, which largely (albeit not completely) follows Bohr's complementarity.[20] In the course of this exposition, Peres makes the following, apparently strange, observation, used as my epigraph for this chapter: "in quantum theory, a particle also has a precise position and a precise momentum" (Peres 2002, p. 27). The observation may, however, not be as strange as it may seem, especially by virtue of its apparent conflict with the uncertainty relations. As I shall argue, Peres's statement, on the contrary, reflects the precise mathematical and physical meaning of the uncertainty relations, and, I shall further argue, as such it also directs our attention toward a new type of relationships between mathematics and physics established in quantum mechanics, especially in the work of Heisenberg and Dirac. It is worth citing Peres's elaboration where the proposition in question occurs more extensively. According to Peres:

> In classical mechanics, a particle has (ideally) a precise position and a precise momentum. We can in principle measure them [i.e., both of them

[20] Peres rightly notes the divergence of various interpretations assembled, along with that of Bohr, under the rubric of the Copenhagen interpretation and also the specificity of Bohr's interpretation. For Peres's more recent treatment of the EPR experiment in the context of information theory, see (Peres 2003), and for his exposition and his own interpretation of quantum mechanics (proceeding more along the lines of quantum information theory), see his book (Peres 1993) and his earlier article (Peres 1984). See also (Fuchs and Peres 2000), which generated considerable controversy, sometimes due to the lack of proper attention to the authors' argument vs. their provocatively polemical title, "Quantum Theory Needs No Interpretation."

> simultaneously] with arbitrary accuracy, and thereby determine their numerical values. In quantum theory, a particle also has a precise position and a precise momentum. However, the latter are mathematically represented by self-adjoint operators in a Hilbert space, not by ordinary numbers. Their [mathematical?] nature is quite different from that of classical position and momentum. In the early quantum literature, operators were called *q*-numbers, while plain numbers were *c*-numbers [according to Dirac]. Likewise, to avoid confusion, we should have used in quantum theory names such as *q*-position and *q*-momentum, while the corresponding classical dynamical variables would have been called *c*-position and *c*-momentum. If such a distinction had been made, it would have helped to prevent much of the present confusion about quantum theory. It is *the imperfect translation* of the *q*-language to the *c*-language that led to the unfortunate introduction of the term "uncertainty" in that context. (Peres 2002, pp. 27-28; emphasis added)

I shall explain my emphasis on "the imperfect translation" presently. Peres, it is true, does not say "a precise position and a precise momentum *at the same time*." If Peres does not mean the simultaneous application of both variables in question in quantum mechanics, the issue is nearly mute—nearly but not altogether, since the question of the relationships between the mathematical and the physical definition of quantum-mechanical variables would still need to be addressed. On the other hand, the parallel with classical physics suggests that Peres does have this simultaneous application in mind, and, as I shall argue, his argument applies, and indeed becomes deeper and more significant in this case. Technically, a proper parallel difference would be in terms of the different character of physical variables—differential functions, subject to differential equations, in classical physics vs. operators in a Hilbert space, subject to certain "algebraic" equations, both of which can be related, but in radically different ways, to real (actually, rational) numbers by means of measurements and predictions. In fact, as explained in Chapter 1, the equations are formally the same, and are thus differential equations, applied to operators as their variables, which is one of the reasons for my quotation marks around algebraic. There are several other reasons for using the quotation marks around algebraic, which I shall address below, in particular that these variables, technically, comprise a noncommutative algebra (rather than a commutative field of functions, as in the case of the variables used in classical mechanics). The main point at the moment is the difference between the variables to which the equations apply, and then the difference in the relationships between these variables and numbers obtained in measurements and (through these variables and equations) in predictions in classical and in quantum physics. These differences are crucial and justify Peres's view expressed in the elaboration just cited.

In particular, in classical physics or, more accurately, mechanics, we can, ideally (Peres is right to qualify), simultaneously assign, as measurable quantities, the values of both variables, that of position and that of momentum, pertaining to the object under investigation, say, a planet in its orbit around the sun. This possibility makes the theory (at least, again, as an idealization) *causal*, insofar as the *behavior* of the objects considered is concerned, and in most cases *deterministic*, insofar as we can *predict* the state of the system under investigation at a future point of time, once we know this state at a given point of time. (As throughout this study, I distinguish "causal" and "deterministic" accordingly.) This possibility also makes the theory realist, insofar as we can phenomenally or mathematically, or both, map the behavior of the system, at least, again, ideally.

In quantum mechanics such deterministic predictions are impossible, in view of the uncertainty relations, regardless of the interpretation of quantum mechanics that one uses. In other words, both the statistical character of our predictions and the uncertainty relations are (correlative) experimental facts. Both, it is worth noting, are also retained even in Bohmian mechanics, realist and causal as the latter is, which indeed allows it to be consistent with the experimental data of quantum mechanics, albeit at the expense of nonlocality, the price that, Bell's theorem more or less tells us, such theories of these data have to pay. I qualify this assessment in view of certain further complexities of Bell's and related theorems, on which I shall comment in the next chapter. In any event, the uncertainty relations make it difficult and in some interpretations, especially those proceeding along the Copenhagen lines, such as that of Bohr, impossible, to assume the behavior of quantum systems to be causal, or to offer any description of this behavior. Hence, one encounters major difficulties with or the impossibility of offering a realist interpretation of the theory. By the same token, the statistical character of our predictions becomes irreducible, again, regardless of the interpretation one uses, even in the case of individual processes or events in question in quantum mechanics. However, by means of procedures such as Born's square-moduli rule, von Neumann' projection postulate, and the like, one can relate each such (operator) variable, but, in view of the uncertainty relations, *never simultaneously both*, to the numerical outcomes of certain measurements that we can perform. One can, ideally, measure or predict *each* such variable with arbitrary precision, within the capacity of our measuring instruments, while the uncertainty relations and, hence, the impossibility to do so for *both* variables would apply regardless of this capacity. As Peres aptly observes elsewhere, "an uncertainty relation [...] is not a statement about the accuracy of our measuring instruments. On the contrary, its derivation assumes the existence of *perfect* instruments" (Peres 1993, p. 93). Bohr, too, notes, in a rather striking sentence: "in this context, we are of course not concerned with a restriction as to the accuracy of measurement, but with a limitation of the well-defined application of space-time concepts and dynamical conservation laws, entailed by the necessary distinction between [classical] measuring instruments and atomic [quantum] objects" (*PWNB* 3, p. 5, also *PWNB* 2, p. 73). Peres's explanation of the uncertainty relations, where this remark occurs, is one of the best available and is consistent with, and clarifies, Bohr's view of them. According to Peres:

> An uncertainty relation such as [$\Delta q \Delta p \cong h$] is not a statement about the accuracy of our measuring instruments. On the contrary, its derivation assumes the existence of *perfect* instruments (the experimental errors due to common laboratory hardware are usually much larger than these quantum uncertainties). The only [available?] correct interpretation of [$\Delta q \Delta p \cong h$] is the following: If the *same* preparation procedure [enabled and defined by the classical control of measuring instruments] is repeated many times, and is followed either by a measurement of [q], or by a measurement of [p], the various results obtained for [q] and for [p] have standard deviations, [Δq] and [Δp], whose product cannot be less than [h]. There is never any question here that a measurement of [q] "disturbs" the value of [p] and vice-versa, as [is] sometimes claimed. These measurements are incompatible, but they are performed on *different* [*quantum objects*] (all of which are identically prepared [with the qualifications given above]) and therefore these measurements cannot disturb each other in any way. An uncertainty relation ... only reflect the randomness of the outcomes of quantum tests. (Peres 1993, p. 93)

This statement also suggests that, their physical, philosophical, and historical significance notwithstanding, the uncertainty relations are, to some degree, a remnant of classical physics, and, from the post-quantum-mechanical perspective, indeed "only reflects the randomness of the outcomes of quantum tests." Richard Feynman expresses this type of view as follows: "I would like to put the uncertainty principle in its historical place. When the revolutionary ideas of quantum physics were first coming out, people still tried to understand them in terms of all-fashioned ideas (such as light goes in straight lines). But at a certain point the old-fashioned ideas would begin to fail, so a warning was developed that said, in effect, 'Your old-fashioned ideas are no damn good when ...' If you get rid of all old-fashioned ideas and instead use the ideas that I'm expounding in these lectures—adding *arrows* for all the ways an event can happen—there is no need for an uncertainty principle!" (Feynman 1988, pp. 56-56, n.3). This procedure of course amounts to estimating probabilities of the outcomes of certain experiments via quantum amplitudes. The situation may be a bit more complex than Feynman makes apparent, but the statement conveys a deep point, which was well understood by Heisenberg and Bohr, and perhaps nothing reflects it stronger experimentally than the impossibility of ever assigning both properties involved to the same quantum object, as here discussed.

It is clear, especially from Peres's last sentence in his elaboration just cited, that, in accordance Bohr's view, the uncertainty relations are here seen as experimentally given, as a law of nature, including as concerns the statistical nature of the tests and phenomena in question, and hence of any theory that would properly cover them. Quantum mechanics does this, since the uncertainty relations can be automatically derived from the formalism. (Any prediction requires some form of a theory to be possible.) It also follows, however, that, in accordance with Bohr's counterargument to most of Einstein's criticisms of quantum mechanics, one cannot, on experimental grounds, in principle, contemplate a physical attribution of both quantities in question to the *same* quantum object, even though one can set up either type of prediction in any given case. There is no experiment that would allow one to make such an attribution, which Einstein thought as possible, with, as will be explained later, the implication that quantum mechanics could be seen as incomplete or else nonlocal. To do so would require two different and mutually incompatible experimental arrangements, in experiments always involving different quantum objects (of the same kind), which, as Peres argues here, entails only statistical estimates concerning the values of a given variable in repeated experiments or of coordination of the relationships between both such variables in the uncertainty relations. I shall return to this point in my discussion of the EPR experiment in the next chapter.

As considered earlier, in Bohr's view, the physical variables themselves, measured or predicted, would pertain to the measuring instruments involved, and such variables can only be rigorously defined by means of actual measurements (i.e., they relate to something that has already occurred and that has been registered in measuring devices), rather than only by means of predictions concerning possible measurements. This last distinction is not significant and indeed does not obtain in classical physics, since all variables involved can be simultaneously defined at any given point. By contrast, in quantum mechanics, we can, on the basis of a measurement, use the formalism to make predictions, in general probabilistic in character, concerning the value of the position variable *associated* with a given quantum object at a given future point in time. (In Bohr's view, all measuring quantities, *c*-numbers, involved would, again, pertain to certain parts of measuring instruments.) If, however, we then, at this future point, precisely measure instead the value of a momentum variable associated with that object, such a measurement would leave the position variable undefined in view of the uncertainty relations or complementarity. For, such an alternative definition would require an experimental arrangement incompatible with that used in the momentum measurement in question. The theory (at least, again, in this type of interpretation) is not realist: it does not assign conventional physical properties or variables, such as position and momentum, to quantum objects themselves and indeed prohibits such an assignment, ultimately, as I noted in Chapter 1, even in the case of a single such variable, whether before, during, or after measurement. Nor, accordingly, is the theory causal or deterministic in predicting the behavior of even individual quantum objects.[21]

Accordingly, in this view, the *physical* definition of any pair of complementary variables can never be simultaneous in this interpretation given the mutual exclusivity of the experimental arrangements which and only which make it possible to assign each variable. (We, again, keep in mind that, in this view, such variables and quantities could only pertain physically to certain parts of measuring instruments, described by means of classical physics, and their definition and assignment is only possible by means of measurements, by registering what has already occurred, rather than by means of predictions alone.) On the other hand, both variables *potentially* involved in one or the other complementary situation of measurement and in the predictions based on such measurements may be viewed as being, at that point, definable *simultaneously* (rather than in a mutually exclusive way) as *formal mathematical entities*. They may, for example, be defined as Hilbert-space operators in the same Hilbert space, where we would consider the commutators corresponding to the uncertainty relations for such variables. This simultaneity, however, does not (at least, again, in this type of interpretation of the quantum-mechanical formalism) correspond to any two events that could actually, physically, occur simultaneously at this point. Indeed, this simultaneity entails, at least, again, in this interpretation, the mutually exclusivity of the two complementary physical events or two pairs of such events, each pair being comprised by a measurement that has already taken place and a possible outcome of a future measurement. That is, this simultaneous definition of both mathematical variables involved entails the impossibility of a simultaneous physical determination or, to begin with, definition of both physical variables involved, concerning which the mathematical formalism in question enables us to make, in turn alternative, predictions. As I said, such predictions are not sufficient to define either variable in any event. In order to do so, at least according to Bohr's ultimate view of the situation, an actual measurement would have to be performed.

In sum, in Bohr's interpretation, the simultaneity of mathematical definition in question does not physically define even each variable by itself, let alone both, for quantum objects themselves, although either variable (but never simultaneously both) variable can be defined, by a measurement, for measuring instruments impacted by

---

[21] Quantum mechanics can be considered as predicting the results of measurement of the values of certain variables with probability equal to unity *in some idealized cases* (which fact was used by EPR in their argument), but not in practice. The question of the statistical nature of quantum processes and events, irreducibly probabilistic even in the case of fundamental individual events rather than only collective or composite events, as would be the case, say, in classical statistical physics, a point stressed by Peres throughout his *Quantum Mechanics* (Peres 1993). I shall, however, bypass these considerations here, since they do not affect the present argument, except by amplifying it. I shall return to some of these questions in Chapter 4.

quantum objects. By contrast, in classical physics, the two simultaneities in question, that of mathematical and that of physical definition of variables, correspond to each other.

This difference may also be shown correlative to the difference, stressed by Bohr and Heisenberg alike from their earliest work on quantum mechanics on, in the role of measurement in classical and in quantum physics. In the case of classical physics, our interference into the behavior of the objects under investigation by means of measuring instruments can, at least in principle, be neglected or compensated for. In the case of quantum physics (at least on a Bohrian view), it cannot even in ideal cases, which fact is correlative both to the uncertainty relations and to the irreducibly statistical nature of the quantum-mechanical predictions. I speak in terms of "interference," since, as Bohr stressed, such expressions as disturbing the behavior of quantum objects by measurements or creating their physical attributes by measurements "are apt to cause confusion" (*PWNB* 2, p. 64). In particular, they may lead to thinking that such objects may independently possess certain physical attributes, which are then disturbed or distorted by the process of measurement. Such is not the case in Bohr's interpretation (it would be in certain other interpretations of quantum mechanics or in Bohmian versions of quantum theory).

Given these facts, the difficulties and imperfections of translation between two situations and languages invoked by Peres, that of classical and that of quantum physics, are not surprising. It was in part and perhaps primarily the prolonged and complex history of quantum theory, from its introduction by Planck in 1900 to the discovery of quantum mechanics in 1925, and many, seemingly insurmountable problems encountered in this history, that were also responsible for these difficulties and imperfections of translation. In a way, this history proceeded by translating (equally imperfectly at first) *c*-numbers into *q*-numbers. The use of the term 'uncertainty,' or 'indeterminacy,' is yet another question, subject to a complex history of its own. Suffice it to say for the moment, that in quantum mechanics our predictions may seen as "determinate," if not "certain," but this (mathematically exact) determination could, in general, only concern the *probabilities* (these, again, determined exactly) of the outcomes of certain future experiments on the basis of the outcomes of certain already performed experiments. Ultimately, one needs to reach much further back in the history of physics, at least to Descartes and Galileo, if not Aristotle, to understand the nature of the translations and confusions in question. Nevertheless, Peres is right to argue that the main source of confusion is a lack of rigorous discrimination in our terminology and in carefully establishing conceptual, physical, and mathematical content of these essentially different situations, that of classical and that of quantum physics. Bohr made related arguments and issued similar warnings throughout his writings.

If we take the differences in question into account, we can easily see that Peres's statement that "in quantum theory, a particle also has a precise position and a precise momentum [at the same time]," that is, a *precisely defined* q-position and q-position, is not at all in conflict with the uncertainty relations. On the contrary, the latter, including their Hilbert space manifestation, $PQ - QP = h/i2\pi$ reflect and indeed establish the precise and, with qualifications given above, even (mathematically) simultaneous definition of both variables in quantum mechanics. The famously noncommutative nature of q-numbers inherent in this formula is itself a crucial part of this precision in definition. This noncommutativity was a pivotal moment for both Heisenberg's and Dirac's work and attitudes, including as concerns the relationships between mathematics and physics in quantum theory, which relationships, I argue here, they radically transformed as a result, as against those that obtain in classical physics.

Peres's article makes the point concerning the definition of variables in quantum mechanics and explains the mathematical, physical, and epistemological character of the theory with an uncommon lucidity and force. Peres is also rightly critical of our persistent sloppiness in commenting on quantum mechanics and on the Copenhagen interpretation, especially Bohr's interpretation. As I said, the article offers a remarkably efficient and illuminating discussion of most key aspects and thorny points of quantum mechanics and of Bohr's interpretation. It is, however, not my main aim here to offer a commentary on Peres's argument, worthy as such an endeavor may be, given our persistent lack of attention to nuances (often more important the more minute they are) in reading this type of arguments. Bohr's writings suffer immensely from this type of sloppiness in reading them, and Peres's discussion of Bohr's views is a welcome exception, also remarkable for its efficiency. My primary aim is, however, different. I would like to explore the implications of Peres's point in question for the relationships between mathematics and physics, which were radically transformed, as against classical physics, with Heisenberg's paper introducing quantum mechanics in 1925, as Bohr, as will be seen, grasped immediately (*PWNB* 1, p. 51). One might argue that these relationships were transformed in part as a result of establishing the conditions for the precise definition of the variables in question, beginning with the noncommutativity of the operator algebra involved.

One can appeal here to either sense of the term algebra, that of the noncommutative operator algebra of quantum 'observables' (may not be the best term, but not much can be done about it now) or the more general sense of the mathematical discipline, especially as opposed to geometry. I shall explain the significance of this general sense of algebra in quantum theory in more detail later. It may be helpful, however, to indicate the main reasons for this significance at this point. Classical physics, both that of the motion of particle-like objects and that of (wave) radiation, may be seen as *geometrical* in the sense of the spatio-temporal character of the behavior of the objects considered in classical physics in our intuitive conception or representation of this behavior. At least such conception or representation is possible in principle.

Accordingly, it possible, for example, to draw actual pictures of such classical behavior, and such pictures were effectively used throughout the history of classical physics from Galileo (who uses such drawings extensively in his works) on, or generate computer images of such behavior. Or, to use the German term prominent in the earlier discussions of quantum mechanics, classical physics is "*anschaulich*," a word usually, albeit imperfectly, translated as visualizable, including by Bohr, who also made the English term one of his key terms and concepts when writing in English. (Bohr also uses such cognates as "pictorial visualization" or "pictorial representation.") By contrast, in quantum mechanics such visualization does not appear to be possible, and is rigorously impossible in Bohr's view and in the view adopted here. It is not, for example, possible to draw a picture of what is actually going on at the quantum level or to generate corresponding computer images, although (not surprisingly) this appears to be possible and have been done in Bohmian mechanics, which, however, is a theory different from the standard quantum mechanics, with which the present study is primarily concerned.[22] Some have tried to make quantum mechanics *anschaulich*, which was one of the reasons for Schrödinger's pursuit of his wave mechanics (and he appeals to the term throughout) or later on for Bohm's version of quantum mechanics. [23] Both were inspired by de Broglie's ideas and philosophy, in part based on the requirement for visualization of the physical processes at the quantum level. One might argue that the success of such projects as rigorous descriptions of the behavior of nature at the quantum level is at best questionable (heuristic use of such "pictures" or even theories is of course a separate question). From the point of view adopted here, this failure itself reflects the new relationships between mathematics and physics in question.

2. TRANSFORMATIONS: FROM GEOMETRY TO ALGEBRA

The preceding argument proceeds from a version of complementarity developed by Bohr in the wake of the argument of Einstein, Podolsky, and Rosen (EPR) and finalized more or less in his "Discussion with Einstein." This version also appears to ground Peres's argument in his article, in part because Popper's critique of quantum mechanics in question is based on an experiment of the EPR type. As discussed in the

---

[22] Bohr's famous illustrations of his arguments, which show heavy stationary measuring instruments (these are, again, described by classical physics, and hence can be represented) and traces of their interactions with quantum objects, but not anything at the quantum level, deliberately represent the difference between classical and quantum physics (*PWNB* 2, pp. 48-49, 54).

[23] For Schrödinger's views of physics and his work on wave mechanics, including as concerns the question of visualization, see volumes 4 and, especially, 5 of Mehra and Rechenberg's *The Historical Development of Quantum Theory* (Mehra and Rechenberg 2005).

preceding chapter, Bohr's post-EPR views retreat from those of the Como version of 1927, which introduced complementarity, and return to Heisenberg's ideas used in developing his matrix version of quantum mechanics. As also discussed there, a number of Bohr's important statements on quantum mechanics occurs in his 1925 survey "Atomic Theory and Mechanics," well before he introduced complementarity in the Como lecture, and before Schrödinger's wave mechanics, but immediately in the wake of Heisenberg's paper introducing his matrix mechanics, a "rational quantum mechanics." The Como version of complementarity was substantially indebted to de Broglie's and Schrödinger's "wave" theories, although not, even at that point, their interpretations of those theories.

Bohr's 1935 reply to Einstein, Podolsky and Rosen was arguably the most decisive work in developing his ideas in the form here considered, although there is earlier evidence of this shift, around 1929, under the impact of his previous exchanges with Einstein. In the wake of the EPR argument, Bohr arrived at a more radical interpretation. This interpretation does not depend on either wave or particle theories or properties, not even in partial terms, in describing quantum objects, to whose characterization and behavior neither description or theory is applicable, as Peres appears to suggest in this article just discussed, via E. C. Kemble's 1937 book, *The Fundamental Principles of Quantum Mechanics*, which he justly credit. The application of such properties to quantum objects was questioned and even abandoned by already in the Como version (*PWNB* 1, pp. 56-57). Some (but not all) properties of either theory are retained at the level of the effects of the quantum, and hence in turn ultimately indescribable, interaction between quantum objects and measuring instruments upon those instruments. This interaction is, in Bohr's language, "irreversibly amplified" to the classical level of these effects (such as traces of the collisions between quantum objects and silver bromide plates), which are physically described in terms of classical physics. Nor, accordingly, would one depend on the wave-particle complementarity, again, never especially favored by Bohr.

As follows from the argument given in Chapter 1, Heisenberg reinvented matrix algebra from physics. The main postulate, the (matrix) multiplication postulate, of Heisenberg's new theory, "and in fact his type of combination is an almost necessary consequence of the [Rydberg-Ritz] frequency combination rules" (van der Waerden 1968, p. 265). "Almost" is an important word here. For, while Heisenberg, to some degree, arrived to this postulate in order to get the combination rules right, through a complex processes of "guessing," or ("guessing" is not the best word here) working with, among other things, the correspondence principle and the data, the justification or derivation is not strictly mathematical. The rule can only be justified by an appeal to experiment, which is the case even for a fully developed matrix (or wave) quantum mechanics (Heisenberg 1930, p. 108). This combination of the particular arrangement of arrays of complex-number elements, linked to the data, and his reinvention, through physics, of multiplication rules for his new kinematical variables, defined by this arrangement, that is, for matrices, was his great and justly famous invention.

It is true that the noncommutativity of Heisenberg's new kinematics may be and often is seen as an equally great, or even greater, discovery, especially in the context of a fully developed quantum-mechanical formalism, which eventually resulted from his initial insights and arguments. Both Born and especially Dirac immediately recognized its fundamental significance and effectively used the feature itself in their work.[24] From the perspective of Heisenberg's process of his discovery of quantum mechanics, however, this noncommutativity is a consequence of his algebraic multiplication postulate (van der Waerden 1968, p. 266). That is, it automatically follows from the character of the new kinematic elements and of the nature of their multiplication, designed to keep the frequency combination rules satisfied. These kinematical elements themselves are in place, however, even for commuting variables, such as the square of a generalized (matrix) coordinate, which (rather than the general case of matrix multiplication) was all that Heisenberg needed in the particular case of the one-dimensional aharmonic oscillator and its Newtonian equation, treated in his paper by his new kinematics and mechanics. He noted the general noncommutativity of the multiplication he introduced for his new kinematical elements in his paper. As he said himself later, he saw this noncommutitative as a potential hindrance, which was threatening the whole project, beyond the particular case that he considered to justify the new mechanics. He was even glad that he did not give up the whole project in view of this unpleasant feature, never encountered in classical mechanics or the old quantum theory. His decision to proceed also made him the discoverer of the precise definition of quantum mechanical variables, such as position and momentum, as matrix elements. One clearly sees this from examining his work leading to his invention of his new mechanics that such a redefinition was his primary concern.

As I said, the classical (e.g., Newtonian, initially used by Heisenberg, or Hamiltonian) equations of motion are formally retained in these relations, but are applied only to matrix variables and no longer to anything describing the motion of particles, which also changes the nature of the equations. The founders of quantum theory used to speak about the algebraic as opposed to differential, or even continuous, nature of these equations. This language is consistent with the view of quantum physics as dealing with discrete rather than continuous magnitudes, but is not altogether accurate. For one thing, Schrödinger's equation, which is a differential equation and was viewed by Schrödinger himself as an alternative to the quantum algebra of Heisenberg's version, is mathematically equivalent to that algebra, as Schrödinger was among the first to demonstrate. Even before Schrödinger's discovery of his equation, Born, Jordan, and

---

[24] It is of some interest that Pauli initially missed the significance of this point, and insisted that these variables should commute in a proper form of the theory.

Heisenberg worked out a form of matrix differential calculus for their matrix mechanics. Ultimately, in both versions, we deal with the more complex and, in general, denser (than in the standard continuum) continuities of Hilbert spaces. The question, correlative to the question of the precision of definition of variables in quantum mechanics, is how these continuities relate to physical nature at the quantum level, in contrast to the continuous representation of physical processes by means of standard differential equations, say, in their Hamiltonian form, in classical physics. On the present view, they do so only in terms of probabilistic predictions of the outcomes of measurements rather than in terms of description of anything at the quantum level. This difference is a crucial part of Heisenberg's revolution, which thus also radically transforms, as against classical physics, the relationships between mathematics and physics.

As explained in this study, according to Bohr's post-EPR view, complementary features of description are defined through the mutual exclusivity of some among experimental arrangements, the possibility of which considered, nevertheless, necessary for a comprehensive description of quantum phenomena, now in Bohr's sense of the term. In such arrangements quantum-mechanical effects manifest themselves—*as macroscopic effects*, pertaining to or manifest in certain (classical) properties of certain parts of measuring instruments—say, a spot left on a photographic plate or a displacement of a certain part of a measuring apparatus under the impact of a quantum object. Each such effect itself can be described in terms of classical physics, but only insofar as we ignore the ultimate processes responsible for the emergence of this effect, that is, the interaction, quantum in nature, between quantum objects and measuring instruments. Once we consider the question of these processes and of this interaction, the situation becomes as follows.

We can observe a change in the state of certain parts of measuring instruments and measure the pertinent classical variables, say, a change in momentum, in a perfectly classical manner, that is, the effect would be the same if the momentum would change in an interaction with a classical object. Or we can use such parts as a classical frame of reference, where we can, say, register the position of a spot left by a "particle" colliding with a silver bromide screen. We cannot, however, account, either classically or quantum-mechanically, for the physical process that led to such changes in the conditions of the measuring instruments, nor use such occurrences for theoretical predictions based on the formalism of classical physics. The necessary classical variables defining the behavior of classical objects in the situations amenable to a proper classical physical description can no longer be all simultaneously (at most only half of them can) assigned to or even be defined for the parts of measuring instruments under the impact of their interactions with quantum objects.

This situation is Bohr's interpretation of Heisenberg's uncertainty relations, now strictly applicable only to certain classically described parts of the measuring instruments, impacted by quantum objects, rather than to quantum objects themselves. In other words, in quantum-mechanical measurement, certain classical variables, namely, conjugate classical variables, pertaining to the key parts of measuring instruments, can never apply simultaneously, since they entail mutually exclusive experimental conditions under which the corresponding complementary phenomena or effects appear. There appears to be no experimental arrangement that would allow us to circumvent this difficulty. This interpretation also gives a proper physical content to, to return to Peres's language, the precisely defined position and momentum measurement in quantum mechanics. That, however, again, means, as Peres makes clear, that both cannot be physically defined simultaneously for the measuring instruments involved and cannot be ascribed, even each by itself, to the quantum objects themselves under consideration (Peres 2002, p. 27). Thus, from this perspective, one may indeed say that the uncertainty relations manifest or are correlative to the possibility of the precise, rigorous definition of conjugate variables in quantum mechanics, such as 'position' and 'momentum.'

In classical mechanics, the possibility of a simultaneous assignment such variables to a given physical object also enables a causal mode of description of the behavior of the objects considered in and making deterministic predictions concerning this behavior, in contrast to, in general, only statistical estimates of the outcomes of experiments, subject to uncertainty relations, in quantum mechanics. In Bohr's interpretation the absence of causality becomes automatic, given that to quantum objects themselves or to their behavior no physical properties of any kind could be assigned. For, as has been noted by Schrödinger, among others, if a physical state cannot be assigned (say, on the model of classical physics) to a quantum system, one could hardly speak of a causal change of the system from one state to another. Thus, Bohr's interpretation of quantum mechanics as complementarity is, in his words, able to respond to "the [experimental] circumstance that, in general, one and the same experimental arrangement may yield different recordings" by arguing that "the logical approach cannot go beyond the deduction of the relative probabilities for the appearance of the individual phenomena under given experimental conditions" (*PWNB* 2, p.73). At the same time, Bohr also notes, "any logical contradiction ... is excluded by the mathematical consistency of the formalism of quantum mechanics, which serves to express the statistical laws holding for observations made under any given set of experimental conditions" (*PWNB* 3, p. 25). These phenomena manifest themselves in our classically described measuring instruments under the impact of quantum objects, which view compels Bohr eventually to define the very concept of phenomenon, as applicable to quantum mechanics, accordingly. The possibility of such a classical description of our measuring instruments also enables us to control the sameness of certain experimental arrangements, as against the impossibility to control the outcome of the events in question in quantum mechanics, the impossibility *in principle*, rather than only in practice, as could be the case in classical physics as well.

Accordingly, on this view, while the mathematical formalism of quantum mechanics defines all variables involved precisely, as precisely as that of classical physics defines its variables, this formalism refers, in general statistically, to the effects of the interactions between quantum objects and the measuring instruments involved upon those instruments and only to such effects. It also predicts, in accordance with the uncertainty relations, the complementary character of some among these effects. By contrast, this formalism does not describe or otherwise account for the physical processes, essentially quantum in nature, that lead to the emergence of such effects. As the mathematical definition of the variables involved (whether we speak of matrix elements, q-numbers, or operators in Hilbert's space) is precise mathematically, this last "definition" is precise physically, at least ideally, and we recall that in classical physics, too, the physical definition of variables is an idealization. The type of idealization used in classical physics appears to be impossible in quantum physics. It may, however, be replaced with a different type of idealization, whereby quantum objects themselves cannot be attributed any physical properties on the model of classical physics, ultimately even single such properties, rather than certain joint properties, as would be prohibited more immediately by the uncertainty relations even in interpretations that are less epistemologically radical than that of Bohr's complementarity. In the case of complementarity, the uncertainty relations, again, apply to certain classical physical variables describing certain parts of measuring instruments impacted by quantum objects. To such parts both complementary variables, such as "position" and "momentum," or the corresponding measurable quantities, cannot simultaneously be assigned. To quantum objects themselves no physical variables or quantities can ever be assigned.

In this type of interpretation, quantum mechanics suspends any possible description and, thus, both realism and (again, automatically) causality at the level of the ultimate objects that it considers—a price, again, too exorbitant and even unacceptable to some, Einstein and Schrödinger, among them. Nevertheless, thus interpreted, it remains a rational physical theory, free from any "underlying mysticism foreign to the spirit of science," in the same sense that classical physics or relativity are, and certainly without presupposing some inaccessible mystical agency (for example, on one or another theological model), responsible for the situation in question in quantum mechanics (*PWNB* 2, p. 63). It conforms to all the standard disciplinary requirements defining modern mathematical sciences of nature, from Galileo on. It provides as rigorous theoretical and experimental knowledge as does classical physics, and it rigorously relates (for example, by means of uncertainty relations) what can and what cannot be known, beginning, again, with the precise mathematical and physical definition of all theoretical quantities involved. It involves experimental verification, logical arguments, the mathematical character of the theory, and so forth. In short, it fulfills all the main requirements of modern physics and modern scientific inquiry in general, not the least insofar as it provides a pathway to the discovery of new physical laws, the primary business of all physics.

Indeed, according to Bohr, beyond being merely consistent with this possibility, complementarity enables quantum mechanics to do so. In other words, complementarity does so not *in spite of* the fact that it suspends realism and causality at the ultimate level, but *because* it suspends them. The reason for this claim is that "it is only the mutual exclusion of any two experimental procedures, permitting the unambiguous definition of complementary physical quantities" (which entails this suspension) that "*provides room* for new physical laws," the laws that quantum mechanics rigorously incorporates in its theoretical framework (*PWNB* 4, p. 80). That is, complementarity and the suspension of reality and causality that it entails at the level of quantum objects themselves allow for the proper place of these laws and of quantum mechanics itself as a physical theory in physics as a modern mathematical science of nature. As Bohr notes on the same occasion, "the coexistence of [these laws] with the basic principles of science [might at first sight appear irreconcilable]" (*PWNB* 4, p. 80). But in fact such is not the case. Complementarity makes this coexistence possible, even if at the cost, to some, again, unacceptable, of reconceiving the nature of physical phenomena and of our way of rigorously treating them. "It is," Bohr contends, "just this entirely new situation as regards the description of physical phenomena, that the notion of *complementarity* aims at characterizing" (*PWNB* 4, p. 80).

3. RELATIONS: BETWEEN MECHANICS AND MATHEMATICS

Although Bohr was to nuance and refine his views still further in subsequent years, the last comment just cited, made in 1935, by and large sums up Bohr's journey from his initial response to Heisenberg's discovery in "Atomic Theory and Mechanics" to his ultimate version of complementarity. Bohr's final comments on Heisenberg's discovery, closing "Atomic Theory and Mechanics," are, however, of considerable interest in this context and might be unexpected, given the subsequent trajectory of his thought, as sketched here, especially his insistence on the crucial role of measurements rather (as would be the case for Heisenberg throughout his life) than on the fundamental role of mathematics. These comments suggest a program rather different from the one he came to follow later, although one might also argue (the present author would) that by taking this view one might underestimate certain subtler and more hidden complexities of Bohr's later views as concerns the significance of mathematics in quantum mechanics. Measuring instruments did, however, come to replace "the mathematical *instruments*," which he invokes on this occasion, in playing "an essential part," *the* essential part, in quantum mechanics in his interpretation, and it would be a curious exercise to translate Bohr's passage as a whole into his later views, accordingly. The emergence of this emphasis in Bohr's thinking is an interesting and important question in its own right, which cannot be pursued here. It is the role of mathematics rather than measurement in quantum mechanics that I want to emphasize, in part against (later) Bohr, although not quite following Heisenberg either, given some of his more Platonist views, since something quite different from Platonism is, in my view, at stake here. On the other hand, it is clear that the question of the role of mathematics in quantum theory and in physics in general continued to remain significant for Bohr's thinking and work throughout his life that it might appears, as is clear from example from his late essay, "Mathematics and Natural Philosophy" (1956) (*PWNB* 4, pp. 155-159). Heisenberg's Platonism or Platonism in general, physical or mathematical, is not a simple matter either, and, on a closer inspection, Heisenberg's view may be closer to the one I am pursuing here than it may appear. Be it as it may, in 1925, in the wake of Heisenberg's discovery, Bohr commented as follows:

> It will interest mathematical circles that the *mathematical instruments* created by the higher algebra *play an essential part* in the rational formulation of the new quantum mechanics. Thus, the general proofs of the conservation theorems in Heisenberg's theory carried out by Born and Jordan are based on the use of the theory of matrices, which go back to Cayley and were developed especially by Hermite. It is to be hoped that a new era of mutual stimulation of mechanics and mathematics commenced. To the physicists it will at first seem deplorable that in atomic problems we have apparently met with such a limitation of our usual means of visualization. This regret will, however, have to give way to thankfulness that mathematics in this field, too, presents us with the tools to prepare the way for further progress. (*PWNB* 1, p. 51; emphasis added)

Bohr proved to be right in anticipating that the situation would be seen as deplorable by many, but he was at best only partially right in his second guess. While the role of mathematics was nearly uniformly appreciated, the return to a classical-like visualization of physical processes became equally desirable and for many, beginning with Schrödinger, a major imperative. Schrödinger's wave-mechanics aimed, first, at a more *anschaulich* wave representation in the phase space and, then, ultimately, at relating the formalism to the space-time (wave) processes at the quantum level. Schrödinger's equation itself can be reinterpreted in accordance with Heisenberg's and then Bohr's views.[25]

---

[25] Heisenberg addressed the question of visualization on several occasions as well, including, importantly, in "The physical content of quantum kinematics and mechanics [*Über den anschaulichen Inhalt der quantentheoretischen Kinematik und Mechanik*]" (Heisenberg 1927, *QTM*, pp. 62-86), introducing the uncertainty relations, as its German title indicates. As I noted in Chapter 1, The English translation of the title is

Ultimately, the mathematical formalism of quantum mechanics manifests a complex and subtle form of algebra, including in its technical sense of "algebras"—matrix algebras, or algebras of Hilbert-space operators, C^*-algebras, and so forth. To these algebraic structures, it is worth noting, the *language* and certain *conceptions* of geometry, such as that of Hilbert *spaces*, and certain forms of geometrical intuition may apply in mathematics and physics alike, as it was, especially, by Dirac and, to some degree, Born. This application of the geometrical language and conceptions is important, in particular, in establishing certain relationships between operators algebras and Hilbert spaces, especially as developed in von Neumann's seminal work. These relationships are manifest, in particular, in Stone's representation theorem and then the theorems of Gelfand and Gelfand, Neimark, and Segal. These theorems enable us to construct a Hilbert space from a given Banach algebra (the "points" of this space correspond to the maximal ideals of this algebra), and hence, as I said, "geometrize" it. This procedure allows one to traffic between different way of representing and handling "observables" (operators) and "quantum states" (state vectors) in the mathematical formalism of quantum mechanics (physical states are, again, a separate issue). These findings also had major implications for algebraic geometry, from its inception in the nineteenth century to Alexandre Grothendieck's work (which uses the same type of construction) and then to Alain Connes' noncommutative geometry introduced in 1980s, which takes Heisenberg's discovery as its main point of departure. Noncommutative geometry itself emerged through various mathematical developments just mentioned and their relationships, with far reaching implications for modern physics, specifically quantum field theory. Via the developments just mentioned and related mathematical theories, such as, among others, K-theory, Galois's theory, algebraic field theory, elliptical curves and Riemann's ζ-function, fractals, and their interrelationships, noncommutative geometry extends to, and was in part developed in conjunction with higher level quantum field theories. This geometry is called noncommutative, in the first place, because, unlike the algebras with commutative multiplication (rings) used in algebraic geometry, the algebras from which the spaces of noncommutative geometry are constructed, are defined by the noncommutative nature of multiplication in them.

It may appear remarkable and is counterintuitive (most aspects of quantum mechanics are) that, in the Hilbert space version of the formalism, the "same" Hilbert space is used to relate, in terms of possible predictions, to *two different and always mutually exclusive actually possible situations*, as opposed to a *single* such situation. (Obviously, using other versions of the formalism would not change the situation, which was indeed analyzed by Bohr in terms other than those of Hilbert spaces.) By the same token, the actual measurement associated with either variable corresponds to different, but

---

misleading and may be better rendered as "on the visualizable [or representable or intuitable] content of quantum-theoretical kinematics and mechanics."

in each case *complete*, *information* about a different state, as opposed to each measurement conveying *partial information* about the same state. One would naturally expect the latter state of affairs to obtain, given the preceding history of (classical) theoretical physics and the use of mathematics there, either in describing the actual physical space on the configuration or phase space, or at least one would have expect this before Heisenberg discovered his new, noncommutative mathematics of quantum theory. But then, a Hilbert space is a very different type of mathematical object from those used in physics hitherto (including in describing the phase space), even in relativity, beginning with the noncommutative nature of the operator algebra in question, or the fact that the Hilbert spaces involved are, *irreducibly*, over complex rather than real numbers. (There is a well-known theorem showing that such must be the case.) This is a tremendous difference already, which was essential for the creation of quantum mechanics in either version, and which, it may be noted, also changed our thinking concerning the very concept of number in mathematics and even beyond.

Peres's article, discussed earlier, has a beautiful epigraph: "I called thee to curse mine enemies, and, behold, thou hast altogether blessed them." I cannot, however, avoid the temptation to note that it comes from *The Book of Numbers* and carries numbers (24:10) with it. While this is undoubtedly merely a coincidence, quantum mechanics, too, is "a new book of numbers" in physics, the book comprised of "expectation catalogues," as Schrödinger called them in his cat paradox paper (Schrödinger 1935, *QTM*, pp. 152-157). One might even argue that this noncommutativity points towards the necessity of complementarity in Bohr's sense, as explained here, including the point, just stated, concerning the completeness of information contained in each complementary measurement.

By the same token, this difference in the mathematical nature of the formalism implies that such notions as the state vector or density matrix need to be thought of essentially differently as concerns their relations to measurement and predictions (in general, again, statistical in nature) from the mathematical objects used for analogous purposes in classical physics, including classical statistical physics. The situations bears significantly on the concept of 'quantum state,' now using the term more generally, including physically and epistemologically (rather than only mathematically as a Hilbert space object) and subject to a number of significant recent exchanges and even controversy around quantum information theory, although the controversy is of course long standing, at least since the introduction of Schrödinger's equation.[26] It is worth addressing the situation here.

If by a 'quantum state' we refer to a state vector or a density operator in a Hilbert space (or a suitable alternative mathematical object, such as a linear form, or the

---

[26] See (Fuchs 2003) for an extensive treatment of the subject and for further references, and (Peres 1984).

algebra of observables associated with a given quantum system, via the representation theorem of Gelfand, Naimark, and Segal), then, from the perspective here outlined, we can roughly see it as follows. A quantum state is part of the mathematical machinery that enables a rigorous (numerical) assessment of our probabilistic expectations concerning the outcomes of certain experiments to be performed on the basis of the data obtained in certain previously performed experiments. Other parts of this machinery include other abstract mathematical elements pertaining to that Hilbert space (or, again, some alternative mathematical machinery) and numerical data, of determinate or probabilistic character, obtained from those previously performed experiments, which also constrain what kind of predictions concerning the objects under investigation we can make. This machinery also supplemented by the rules, in general independent and not derivable from the formalism itself, and introduced *ad hoc*, such as Born's square-moduli rule, by means of which the formalism and the data could be linked to numerical statistical predictions. The crux of the question is whether, to what degree, and in what sense such mathematical states correspond to actual physical states of the quantum systems considered, for example, in the way they do in classical physics, in which the concept of (physical) state originates and from which it initially comes to quantum mechanics. In the view adopted here, the answer to this question is negative: quantum states do not correspond to any physical states of quantum objects or of anything else (specifically measuring instruments) at all. On the other hand, closer to recent arguments in quantum information theory, quantum states can be seen as "mental" (not the same as mental states!). Or more accurately, these states coupled to the rules for deriving from them the probabilities of the outcomes of the experiments in question, relate or even correspond to our probabilistic expectations concerning these experiments. Such expectations are always mental. The probabilistic data defining the expectations can of course physically exist otherwise, say, as written down somewhere after the corresponding measurements are recorded and the corresponding calculations are performed. It is nevertheless meaningless to speak of such expectations apart from our mental processing of them and hence apart from somebody holding such expectations in one's mind (possibly in one's unconscious).

In some respects, this is true in classical physics as well. In that case, however, we can, at least in principle and in idealized models, rigorously correlate such states of mental expectations and physical states on the system in question, and also, at least in principle, disregard or compensate for the role of the measuring instruments deployed. Neither this type of correlation nor this type of compensation is possible in quantum mechanics in the present interpretation. The irreducible role of measuring instruments makes quantum states strictly a mathematical tool of our mental expectations. They do not relate either to the motion of quantum objects themselves or any properties of these objects at the time of their interactions with measuring instruments (e.g., collisions with the screen in the double-slit experiment), with the outcome of which interactions the expectations in question are concerned. Instead 'quantum states' are used to relate to measurements and predictions concerning observations pertaining strictly to certain parts of measuring instruments impacted by their interaction with quantum objects. In other words, quantum states relate physically to no properties of quantum objects at any point—before, during, or after their interaction with measuring instruments. They only relate, in terms of predictions, to certain classical physical states of systems (measuring instruments) described by means of classical physics, without, again, in any way relating to any physical processes, classical or quantum, responsible for these classical states. There is no physical evolution of any kind or at any level, quantum or classical, to which quantum states relate. They do, however, again, accompanied by certain *ad hoc* rules, enable us to make excellent (statistical) predictions concerning the outcome of the interactions between quantum objects and measuring instruments as these interactions manifest themselves in the classical physical properties of those instruments.

The mathematics of quantum theory, quantum states, as mathematical objects, included, enables the predictions in question, but, at the same time, it describes nothing physically involved in the situation. By contrast, while classical physics can be used to physically describe the properties of measuring instruments manifesting the effects in question, it cannot account for these effects in its usual manner, that is, by mathematically describing the behavior of *quantum objects* and predicting the effects in question in terms of properties of these objects. Nor can it make correct predictions concerning the effects in question. I here refer of course to the theories of classical physics, from classical mechanics to classical statistical physics and classical electrodynamics, rather than to epistemologically classical-like quantum theories, such as those of the Bohmian type. The (complex) Hilbert space offers us an extraordinary mathematical tool to do quantum theory, but, it appears, it also makes this theory irreducibly Bohrian epistemologically.

A Hilbert space used in any quantum-mechanical situation is not a physical space in any conceivable sense, nor is it even anything like (the closest analogy) the phase space of classical physics, including in classical statistical physics, even though the latter is geometrically quite different from the classical (Euclidean) physical space used in classical mechanics. (I leave relativity aside for the moment.) Schrödinger was clearly aware of this difference from the outset of his work on wave mechanics, and he indeed started with the idea of developing a more *anschaulich* treatment of the phase (or configuration) space in mind, in part as against de Broglie's more space-time picture of matter waves. The phase space of classical physics, in a realist, at least a reasonably realist, manner, allows us to connect itself to the physical space, via (commuting) conjugate (function) variables of classical physics, as opposed to (noncommuting) conjugate (operators) variables of quantum mechanics. One indeed wonders whether we should call these operators "position" and "momentum" operators, let alone "observables," any more than we can call a (complex) Hilbert space a "space," or its elements "vectors," except by analogy and metaphorically. On the other hand, this mathematics does work in allowing us to precisely define them as variables and enables the functioning of quantum mechanics as, to return to Galileo's language, "a mathematical science of nature," and, reciprocally, allows us to rethink the very idea of space in mathematics and physics, and their relationships.[27] These questions would arise even in the case of the two-dimensional Hilbert spaces that we use in working with spin, which is not a "spin" either, let alone infinite-dimensional Hilbert spaces, which we need in working with continuous variables, such as "position" and "momentum." (One can remove the quotation marks if these terms are seen, as in Bohr, as denoting classical variables pertaining to measuring instruments impacted by quantum objects, variables that, as explained earlier, are, however, subject to the uncertainty relations, as opposed to those of classical physics.)

In short, we deal with a very different type of relationships between these mathematical entities and the physical experimental data than in classical physics, including as concerns the question of probabilistic predictions. That is, the character of probability would be affected accordingly, at least its physical character (i.e., the nature of the physical situation responsible for probability in quantum mechanics vs. that of classical statistical physics or that of Bohmian theories). As noted earlier and as will be seen in detail later, it automatically follows that, in contrast to classical mechanics, in quantum mechanics we must, in general, restrict ourselves to statistical predictions even when dealing with individual, rather than only collective, events.

Accordingly, quantum mechanics compels us to think of the radically new relationships between mathematics and physics, and between both and nature, which argument, as we have seen, in fact follows that of Bohr, who, at least in 1925, saw this possibility in Heisenberg's matrix mechanics (*PWNB* 1, p. 51). These relationships are altogether different from the program of the mathematization of nature, which so beautifully and effectively shaped classical physics and, somewhat differently, relativity. That program may also be seen as geometrical, the view that lies at the origins of modern physics in Galileo and Newton, who we recall was compelled to present his mechanics in terms of Euclidean geometry rather than calculus. For Galileo, physics was a subset of Euclidean geometry, defined (in our present terms) by virtue of the introduction of mass, and it was presented in his work in terms of geometrical figures, as, importantly, idealized representations of nature. Some algebra was inevitably involved as well, as was of course the case in Newton or Descartes, helped by his invention of analytic geometry, a highly relevant history, which cannot, however, be pursued here.[28] In some respects,

---

[27] For a suggestive and far-reaching meditation on this subject, see Pierre Cartier's article (Cartier 2001) or, again, Alain Connes' work (Connes 1994), upon which Cartier's argument is partly based.

[28] For a discussion of Galileo in this context, see (Plotnitsky and Reed 2001).

Schrödinger's program of his *anschaulich* wave mechanics was, along with Einstein's relativity and de Broglie's wave theory, among the attempts to follow this type of program (forever longed for by Schrödinger), which Bohr's earlier quantum theory and then Heisenberg's matrix mechanics appeared to have made all but impossible. Indeed, Heisenberg's work and the subsequent developments of quantum theory, from quantum electrodynamics on, appear to tell us that such a mathematization may be rigorously impossible at the quantum level. This impossibility would make quantum mechanics an essentially algebraic and, in Bohr's terms, essentially *symbolic* theory. From this viewpoint the mathematical nature of quantum theory is just as essential as the irreducible role of measurement in it, perhaps ultimately more so. This contention may go against the letter of Bohr's understanding and interpretation of quantum mechanics, although I do not think that it actually does. In any event, it would still be in the spirit of Bohr, at least at that defining moment in 1925, when he confronted quantum mechanics, introduced by Heisenberg, as a rational and precise theory of quantum phenomena and when he realized, rightly, that it would actually become one. He also rightly saw the new significance of mathematics in quantum theory, and hence the new relationships between mathematics and physics.

This character of the theory would also essentially affect the practice of theoretical physics in the quantum domain. I would argue that this is precisely what occurred, beginning with Heisenberg's and then (and perhaps especially) Dirac's work, whatever the philosophical attitudes of the practitioners themselves may be, for example, as concerns the possible visualization of physical processes or their possibly non-visualizable representation by the formalism (as opposed to merely using it in its predictive capacity). I am, let me reiterate, speaking of the essential significance here, since visual intuition concerning physical processes, including on quantum level, may play an important, if, I would argue, inevitably partial, role, in the actual thinking of physicists, the communication of ideas and findings, and so forth. In this "algebraic" paradigm, however, the practice of theoretical physics is transformed into working with the mathematical apparatus of the theory (while building upon the preceding mathematical architecture) to make this apparatus enable correct predictions, rather than trying to develop an idealized mathematical description of the physical processes considered. Dirac famously spoke, including in describing his discovery of his even more famous equation for the (free) relativistic electron, of most of his work as "playing with equations," to which expression I want to give a somewhat more rigorous meaning here (Dirac 1962). This "playing with equations," as a mathematical and, in this sense, algebraic … it is not easy to find the right word here … *work* seems to be most fitting, is then related to the outcome experiments in terms of actual numbers, by applying these equations to the numerical data obtained in the previously performed experiments. In this sense, Bohr's view of quantum theory as a set of "algorithms" (his term) for predicting, inevitably statistically, the outcomes of certain experiments to be performed on the basis of the outcomes of the previously performed experiments inevitably involves and is based in this practice, which has, however, indeed as a consequence of a greater complexity, rigor, and precision. This is in particular how Dirac discovered his equation. This is what his "playing with equations" was. A very serious business indeed!

A precise definition of quantum-mechanical variables largely emerged from this type of practice of theoretical physics, and may be impossible otherwise. For one thing, in contrast to classical physics, it may not ever be possible (it is not in the interpretation here considered) to observe or measure anything pertaining to quantum objects themselves or their behavior, and then use these data in our theories of this behavior even by way of idealized models of the classical type.

The theoretical work just described is not quite the same as mathematics, specifically as concerns the use of (rigorous) mathematical proof, and indeed sometimes involves a manipulation of mathematically illegitimate objects, which may (such as Dirac's delta function) or may not be eventually made mathematically legitimate, but it is mathematical nevertheless, as many (but not all) mathematicians came to recognize.[29] At the same time, however, and by the same token, these relationships make mathematics ever more indispensable for the practice of physics, as a precise mathematical science of nature, including in discovering new laws of nature, the primary task of all physics. We may, it is sometimes observed, be lucky to have a solar system, visible to our naked eye, as a model on which, suitably idealized, we were able to build classical physics. If to observe the Moon we would have to shoot Jupiter into it and study the debris (a rough equivalent of the experiments in quantum physics), the situation would be quite different. Nevertheless, we may be luckier still, it is almost a miracle indeed, to have the mathematics and the precision, mathematical and physical, of quantum mechanics, where such a model cannot apply, in part because there may be nothing we can see in this case, however powerful our physical instruments are. Our mathematical instruments are luckily precise enough, at least so far.

---

[29] On this point, see (Cartier 2001). The work of Heisenberg and Dirac, and of course earlier that of Einstein, provided major stimuli for modern mathematics and led to many important mathematical discoveries. However, awarding, a few years ago, the Fields medal (at the time the highest prize in mathematics) to Edward Witten, whose work on, among other things, string theory, was a shaping influence on several major recent developments in mathematics, but who has, technically, never proved a single mathematical theorem, caused considerable controversy and consternation on the part of some mathematicians. Hermann Weyl has perhaps unmatched credentials in the twentieth century in combining mathematics and physics in his work, but he has his precursors as well in, among others, Karl Friedrich Gauss and Bernhard Riemann, or of course Newton and Galileo (although in their time one could hardly speak of the difference between mathematics and physics).

In his book *The Continuum*, Hermann Weyl made the following profound observation: “the conceptual world of mathematics is so foreign to what the intuitive continuum presents to us that the demand for coincidence between the two must be dismissed as absurd. Nevertheless, those abstract schemata supplied us by mathematics must underlie the exact science of domains of objects in which continua play a role” (Weyl 1918, p. 108). The situation acquires a special poignancy in quantum mechanics, leading to the situation here considered, where the general (phenomenal) intuition of physical processes has no essential role at all (although it may, again, play an important auxiliary role in physicists’ thinking). Quantum theory, however, involves and depends, and helps us, including mathematicians, to develop our mathematical intuition, whether concerning continuous or discontinuous objects. Hence, it also helps the precision of our mathematical and mathematically physical theories and the definitions they involve, for example, both a precisely defined position and a precisely defined momentum in quantum mechanics, in spite and indeed because of the uncertainty relations.

Chapter 3. Complementarity, Quantum Entanglement, and Locality

1. "THE PECULIAR INDIVIDUALITY OF QUANTUM EFFECTS"

The current state of the debate about quantum mechanics is dominated by the arguments concerning the Einstein-Podolsky-Rosen (EPR) type experiments and "quantum entanglement," the existence of a particular type of correlation between certain spatially separated quantum-mechanical events. This particular way of thinking about the famously strange character of quantum physics emerged in the wake of Einstein, Podolsky, and Rosen's article, "Can Quantum-Mechanical Description of Physical Reality be Considered Complete?" (EPR 1935) and Bohr's reply under the same title (Bohr 1935). While quantum entanglement and correlations, sometimes referred to as "the EPR correlations," were de facto introduced by EPR or, in any event, implied by the EPR experiment (EPR did not consider the question of correlations as such), the shift of focus towards this problematic in the debates concerning quantum mechanics is largely due to subsequent developments. Most notable among them are Schrödinger's "cat-paradox" paper (1935), which introduced the term entanglement [*Verschränkung*], David Bohm's reformulation of the thought experiment proposed by EPR in their paper in terms of spin, John Bell's and related theorems and special cases of quantum entanglement (which expressly deal with correlations), Alain Aspect's experiments, and more recently quantum information theory and related experimental work, such as that of Anton Zeilinger and his group, as well as recent investigations of decoherence and related questions. Einstein's arguments subsequent to EPR's article refined the EPR argument by more sharply focusing it on the possible nonlocality of quantum mechanics (rather than on its incompleteness as a physical theory) and proposed an alternative between seeing quantum mechanics either as a complete but nonlocal or a local but incomplete theory.[30] This

[30] "Nonlocality" will to be understood here as the possibility of physical influences between spatially separated event that would be in conflict with relativity. Other terms, such as "separability," and different definitions of "locality" are sometimes used as well, but these variations and differences are not germane to and may be correlated with my argument. The terms "nonlocality" is sometimes invoked without implying a violation of the Lorentz invariance, although the readers do not always carefully follow these nuances, and the authors do not always adequately explain or qualify them. In this study, unless stated otherwise, nonlocality (and references to it elsewhere, as, for example, in Einstein, Bohm, and Stapp) is used strictly in the sense of the lack of the Lorentz invariance and incompatibility with relativity. Entanglement is used otherwise. I shall also distinguish *the EPR experiment*, by which I mean the

alternative is briefly suggested at the end of EPR's article as well, but is not adequately developed there, and in Einstein's earlier arguments. The question of locality is the main focus of the developments just mentioned, and to some the EPR correlations indicate the nonlocality of either quantum mechanics (whether it is seen as a complete theory or not) or possibly quantum phenomena themselves.

By contrast, Bohr views the completeness of quantum mechanics as fully compatible with its locality. At least Bohr offers an interpretation of quantum mechanics, complementarity, which, while preserving quantum entanglement, is local. It follows that only entanglement is inevitable in quantum mechanics and may, accordingly, be seen as part of the quantum-mechanical data or a form of quantum phenomena, while nonlocality is not. As I argue in this study, compelling reasons for Bohr's epistemology emerged from the outset of quantum theory, in particular those pertaining to certain features of quantum-mechanical formalism and, correlatively, to the question concerning what physical events in space and time quantum-mechanical predictions may refer to. The requirements of locality provided further support to his views, even though Einstein's critique made Bohr refine, sometimes significantly, his argument. Bohr, however, was in possession of a framework that enabled him to address these additional complexities and to refine complementarity accordingly.

The literature on the question of quantum entanglement is massive, and it would not be possible and for my purposes not necessary to address it here, beyond a few works, especially pertinent to my argument in this study. While, however, this literature confirms the significance of Bohr's thought for our understanding of quantum theory and its interpretation, it also reflects persistent problems of reading Bohr's reply to EPR and his writing in general, beginning with Einstein and Schrödinger, and extending to Bell and beyond.[31] My aim here, in accordance with the general approach adopted in this study, is to offer a reading of Bohr's reply and subsequent writings in order to help to

---

thought experiment they propose, and *EPR's argument*, by which I mean their analysis of this experiment and the conclusions they derive. This analysis and, correspondingly, these conclusions are questioned in *Bohr's counterargument*, which offers a different analysis of the EPR experiment and derives different conclusions concerning its meaning.

[31] See especially the commentary by J. Bell (Bell 1987, pp. 155-56, 189-90) and H. Stapp (Stapp 1989, pp.162). Bohr's views and writing encounter considerable resistance among the proponents of Bohm's approach, who find their inspiration in Einstein's and Bell's critique, as, to give two recent prominent examples, in (Cushing 1994) and (Beller 1999). To paraphrase Bohr on EPR's argument, the trends of the latter arguments do not seem to me adequately to meet the case Bohr's writings present to us. Bohm himself had a more balanced view and cogent sense of Bohr's argument and work, which he admired (Bohm 1995). For arguably the final version of Bohm's theory (originally introduced by Bohr in 1952), see (Bohm and Hiley 1993).

establish a better platform for our reading of Bohr, the enormous labor that cannot be accomplished by a single chapter or even a single study.

As much of this study, my argument in this chapter proceeds from "Discussion with Einstein," which, as I said, was originally published in 1949 in the "Schilpp volume," honoring Einstein (Schilpp 1949). Historically, it might be more appropriate to consider first Bohr's reply to EPR, which, along with related arguments by Einstein, is, I argue, is the most decisive work in developing Bohr's ideas in their ultimate form, which is my main concern here. My approach, however, is guided by Bohr's own view of his argument concerning the subject, according to which "Discussion with Einstein" elucidates and, in some respects, refines Bohr's response to Einstein's arguments and "give[s] a clearer impression of the necessity of a radical revision of basic principles for physical explanation" (*PWNB* 2, p. 61). Bohr felt this necessity throughout the history of quantum theory, but it became especially pressing for him in the wake of the EPR argument. "Discussion with Einstein" becomes Bohr's most comprehensive and consistent exposition of his interpretation of quantum mechanics. It also explicates and sometimes reinterprets his earlier argument, those concerning the EPR experiment included.

Nevertheless, a proper understanding of the key ideas of the article requires taking into account these earlier works, beginning with the Como Lecture, "The Quantum Postulate and the Recent Development of Atomic Theory," which introduced complementarity. Although somewhat different in its approach (one might say that it is more "philosophical," while Bohr' post-EPR argumentation is more "empirical") and in some of its epistemological elements, the Como lecture remains germane to Bohr's thought and puts in place most key ingredients of Bohr's post-EPR version(s) of complementarity. In particular, Bohr's argument, in "Discussion with Einstein" and other late essays, concerning "the peculiar individuality of quantum effects," crucial to his epistemology, is a refinement of his argument in the Como lecture. The latter presents its argument in terms of Planck's constant *h*—Planck's quantum of action—and "the quantum postulate," rather than in terms of "effects" and "phenomena," as in "Discussion with Einstein" and other latter works, which concepts, as we have seen, also enable Bohr to rethink the concept of the quantum postulate. The Como lecture introduces the subject of and the very term "individuality" (although not yet of "effects") by way of a hesitant parenthesis. Bohr says: "Notwithstanding the difficulties which … are involved in the formulation of quantum theory it seems that its essence may be expressed in the so-called quantum postulate, which attributes to any atomic process an essential discontinuity, *or rather individuality*, completely foreign to classical physics and symbolized by the Planck's quantum of action" (*PWNB* 1, p. 53; emphasis added). This seeming unassuming shift from "discontinuity" to "individuality" (eventually further supplemented by "indivisibility" and "wholeness") is in fact the opening move in the development of one of Bohr's most radical and innovative concepts.

It took him twenty years and much further development of his interpretation of quantum mechanics before this concept crystallized at the time of "Discussion with Einstein." By then it came to designate phenomena that appear only at the level of the interaction, irreducible in Bohr's interpretation, between quantum objects and measuring instruments, rather than anything applicable to quantum objects themselves. As we have seen, the very concept of phenomenon as applicable in quantum mechanics was by then redefined by Bohr in the manner correlative to his view of "individual quantum effects." Indeed both are equivalent, and further lead Bohr to his concept of atomicity. By the time of "Discussion with Einstein," Bohr argues that "the peculiar individuality of quantum effects presents us, as regards the comprehension of well defined evidence, with a novel situation unforeseen in classical physics and irreconcilable with conventional ideas suited for our orientation and adjustment to ordinary experience [on which classical physics is based]. It is in this respect that quantum theory has called for a renewed revision of the foundations for the unambiguous use of elementary concepts" (*PWNB* 2, p. 62). The question of quantum entanglement and locality of quantum mechanics contributed greatly to Bohr's life-long thinking concerning complementarity and, as I said, in some respects made these revisions necessary.

What is this "individuality" and what makes it so peculiar, and why "effects" (a term that itself became essential for Bohr and persistent in his later writing), rather than quantum objects themselves? The lineaments of Bohr's answers to these questions must be apparent from the preceding discussion of Bohr's epistemology. Accordingly, I shall only revisit some of the key point of this discussion here, especially pertinent to the questions of completeness and locality of quantum mechanics, and to my analysis of the EPR argument and Bohr's reply, which are my primary concerns in this chapter.

Bohr sees the quantum-mechanical situation as indicating (the phrase defines his critique of EPR's argument) *the essential ambiguity* in ascribing conventional (and ultimately any) physical attributes, such as wave-like or particle-like space-time behavior (or kinematical and dynamical variables of classical physics), to quantum objects themselves or in referring to their independent behavior. According to Bohr, any attempt subdividing quantum phenomena will only introduce "new possibilities of the interaction between [quantum] objects and measuring instruments" which "reveal the ambiguity in ascribing customary physical attributes to atomic objects" (*PWNB* 2, pp. 40, 51). Accordingly, any analysis of quantum objects and processes themselves becomes "*in principle* excluded" and is replaced with an analysis of phenomena in Bohr sense, introduced in response to this situation and defined by referring exclusively to the effects of the interaction between quantum objects and measuring instruments under specified experimental conditions. As a result, the undesirable or paradoxical features of quantum phenomena, such as those of the double-slit experiment or those the EPR experiment, which unavoidable appear when one refers to the properties of quantum objects, become removed without affecting the integrity of the data or formalism of quantum theory. This removal is, as we have seen, possible by virtue an immediate application of the standard logical deduction under hardly standard and to some (Einstein, in particular) indeed outright unacceptable, but far from illogical or irrational, epistemological assumptions entailed by Bohr's interpretation. More generally, Bohr's interpretation is defined by its deliberate logical consistency; and Bohr specifically denies that any departure from ordinary logic (for example, multivalued logic) is necessary, at least within his interpretation (*PWNB* 3, pp. 5-6). Along these lines (there are considerable differences otherwise), Bohr's interpretation may be related to Omnés's "logical interpretation of quantum mechanics" (Omnés 1994, Omnés 1999).

It is also worth recalling that, in Bohr's ultimate interpretation, a reference to quantum objects or processes themselves would remain ambiguous even when one speaks of single such attributes, rather than, as would be more common, of a simultaneous attribution of joint properties, involved in various uncertainty relations, and even at the time of measurement. This view appears to have emerged as part of Bohr's thinking concerning the EPR experiment, even beyond his reply, around the time of Warsaw lecture, alongside his concept of phenomenon. His reply, however, clearly contains intimations of this view. As he says there, "In fact to measure the position of one of the particles can mean *nothing else* than to establish a correlation between its *behavior* [not its position!] and some instrument [...] which defines the space frame of reference" (*PWNB* 4, p. 79). Ultimately, this correlation is deprives of its quantum-level correlata, which would allow us to ascribe at least a position to a quantum objects.

In this view the mutually exclusive measurements of either the position or the momentum involved in the uncertainty relations could only apply to the corresponding part of the measuring instruments involved impacted by quantum objects. A given arrangement can only be seen as suitable either for one measurement or another, but never for both together, and even considering both such variables, for example, in order to meaningfully relate them in a given uncertainty relation, would indeed inevitably require at least two experiments and, hence, two quantum objects. In other words, in terms of the corresponding variables of the measuring instruments involved, both variables can never be defined simultaneously and, hence, in a single experiment. As applicable to quantum objects, not even a single variable can be defined. As a result the statistical considerations or, to return to Peres's language, the randomness of quantum tests, become an irreducible part of the overall picture.

These circumstances become essential to Bohr's arguments with Einstein, specifically those concerns the EPR-type experiments. For in this case, too, we are not dealing with, as Einstein called them, "elements of reality" pertaining to quantum objects (here, a "position"), such as those that EPR want to attribute to them, but only with the (in terms of their physical description) classical properties pertaining to measuring instruments or the reference frame they establish. The former type of reference, or at least an assumption of its unambiguous possibility, is found in most of Einstein's arguments, and is responsible for the essential ambiguity of these arguments. The very essence of complementarity, as an interpretation of quantum phenomena, is this mutual exclusivity of two distinct experimental arrangements in such situations, including that of the EPR experiment, as opposed to dealing with a single experimental arrangement or, to begin with, a single underlying (quantum) system, from which different "elements of reality" could be arbitrarily extracted. What, according to Bohr, Einstein did not realize or at least did not fully appreciate and what "we *must* realize" (emphasis added) is "that in the problem in question we are not dealing with a *single* specified experimental arrangement" or phenomenon, from which we can extract one or another element of reality at will, as EPR deem possible. Instead, "we are referring to *two* different, mutually exclusive arrangements" and, accordingly, phenomena (*PWNB* 2, p. 57; also *PWNB* 4, p. 78). Indeed, this point may be seen as reflecting an experimental fact rather than merely an interpretation of quantum phenomena (now using the term in its conventional sense of referring to what is actually observed), as opposed to Bohr's concept of phenomena, which is *an* interpretation of these phenomena. In other words we are not, or need not be, in the freedom of choice with respect to a given single physical situation or state of reality (at the level of quantum objects) as concerns two different elements that this reality possessed, as Einstein appears to have thought, which view of the situation may indeed entail nonlocality. Instead we may see the situation as offering us the freedom of choice in dealing with *two different situations of physical reality* at the macro-level, ultimately, without referring to the attributes of quantum objects themselves (*PWNB* 4, p. 78).

This view leads to an epistemology and an interpretation of quantum mechanics that allows for locality, even as it preserves quantum entanglement, along with all other standard quantum-mechanical data. Quantum mechanics predicts the numerical effects involved and does so in any interpretation, the fact that, as will be seen, is actually used by EPR. (In general, one does of course need one theory or another to make such predictions.) In Bohr's interpretation it *only* enables such predictions, without offering any description or even any reference to quantum objects themselves or their behavior. Indeed, in this interpretation, such a reference is "*in principle* excluded," and the concept of reality itself can only apply at the level of phenomena rather than quantum objects themselves. The latter, again, may be real in the sense that they exist but appear to prevent us from conceiving of the way in which they exist.

As discussed in Chapter 1, quantum individuality itself is no longer seen in terms of quantum objects or processes themselves, but in terms of individual phenomena in Bohr's sense, compelling Bohr to speak of a new feature of atomicity, foreign to the mechanical (atomic) conception of nature, on which classical physics is based. The latter would describe the behavior of its objects as such (in, at bottom, causal, even if practically statistical or chaos-theoretical, terms). By contrast, in Bohr's interpretation, the only ultimate "atomic" (i.e., indivisible) entities that can be rigorously described by quantum theory are certain indivisible configurations of experimental technology, phenomena in Bohr's sense. Any attempt at further subdividing such phenomena can only produce other phenomena of the same indivisible nature, other "atoms," without, as noted above, even making its possible to reveal any property of quantum objects and processes. In other words, no description of or even conception concerning quantum objects is possible, regardless of how far we can divide and subdivide our experimental technology or matter itself.

Thus, contrary to Einstein's view, quantum mechanics in Bohr's interpretation allows for an interpretation that does "offer an *exhaustive* description of the *individual* phenomena" rather than "merely ... the means of accounting for the average behavior of a large number of atomic systems [as statistical ensembles]" (*PWNB* 2, p. 61, emphasis added). But it redefines what "individual phenomena" are or, in Bohr's interpretation, can possibly be, given the data in question in quantum theory. Complementary phenomena can never be reduced to or derived from a single experimental arrangement, or, again, to the possibility of extracting the values of such quantities from a single arrangement, especially as referring to the underlying (independent) quantum object(s). There is no *single* physical situation of which any two complementary physical situations could be parts or aspects, or from which they could be derived.

Einstein said (in 1936) that "to believe this [i.e., that quantum theory offers an exhaustive description of individual quantum phenomena] is logically possible without contradiction" (Einstein 1936, cited in *PWNB* 2, p. 61). Indeed, he thought that Bohr did most justice to the problem, in his reply in the Schilpp volume. The latter also contains an exposition of his own views on the subject (Einstein 1949). However, he saw it "as so very contrary to [his] scientific instinct that [he could not] forgo the search for a more *complete* conception" (Einstein 1936, cited in *PWNB* 2, p. 61, emphasis added). By this point, Einstein saw the situation in terms of the alternative for quantum theory as that between being "complete and nonlocal" vs. "local but incomplete," and appears to (mis)read Bohr along these lines, in part, I would argue, because his missed Bohr's concept of phenomenon as here explained. Bohr argues, however, that a local and complete interpretation of quantum mechanics is possible, once one accepts his epistemology, which is based on this concept and according to which "a more detailed analysis of atomic phenomena [...] is, *in principle,* excluded."

As explained in the Introduction, I here maintain the contextualization of this argument as pertaining to Bohr's interpretation quantum phenomena and quantum mechanics, even if Bohr had a stronger, less interpretation-dependent, claim in mind. He might have thought that this claim inevitably follows from the formalism of quantum mechanics and perhaps (along with the relevant aspects of the formalism itself) from quantum phenomena and their features, specifically the uncertainty relations. He might have been right. A more limited claim, however, suffices for my point at the moment and my main argument in this chapter and this study, which primarily concerns the *possibility* of a local and complete interpretation of quantum mechanics or to begin with, a local interpretation of quantum phenomena themselves and the fact that Bohr's fulfills both requirements. From this perspective, a sufficient counter-argument to that of Einstein is defined by the view just formulated: once the epistemology in question is accepted, it becomes possible to interpret quantum mechanics as both complete and local. Of course, as Bohr noted on the occasion just cited, Einstein's "attitude, even if [...] [it] might seem balanced in itself" still "*implies* a rejection of [Bohr's] argumentation" on philosophical grounds, even if Einstein saw the latter argumentation as compatible with locality, which is, again, not altogether clear (*PWNB* 2, p. 62; emphasis added). The epistemological cost appears to have been as exorbitant or even unacceptable to Einstein as that of nonlocality, and he did not believe that nature is likely to extract this cost from us. Would, however, a lack of conformity with certain classical-like models, whose ultimate correspondence, however indirect and mediated (Einstein's position on this was complex), to "physical reality" has never been established, be enough to see Bohr's conception as lacking in completeness? Bohr certainly did not think so, although he in turn admitted that Einstein's "attitude might seem balanced in itself" (*PWNB* 2, p. 62). In any event Bohr's interpretation is at least sufficient, if not inevitable, to argue for both locality and completeness of quantum mechanics, or the locality of quantum phenomena themselves. No experimental conflict with and no local alternatives to his interpretation, were on the horizon at the time, and, as I would argue, in essence (there are nuances to be considered) not even now.

2. FORMALISM, PHENOMENA, AND THE "CUT"

As stressed from the outset of this study and as the preceding discussion in this chapter makes clear, in Bohr's interpretation, the formalism of quantum theory refers, in terms of in general statistical predictions, only to individual quantum effects rather than to properties of quantum objects or features of quantum processes (in space and time) themselves. In other words, while certain ("effect"-constituting) parts of measuring instruments are described by the formalism of classical physics, there is no presupposition that the quantum-mechanical formalism in any way describes the ("undisturbed") quantum process before the measurement interference takes place, or between instances of such interference. Instead this formalism predicts the outcomes of these interactions and as concerns their impact on the measuring instruments. Hence, it would not be possible to speak in terms of the physical reality of, say, states in linear ("coherent") superposition in the sense that the corresponding formalism (such as Schrödinger's equation or the Hilbert-space formalism) describes the physical (in space and time) behavior of quantum objects themselves. The formalism instead is, again, seen as predicting particular effects pertaining to the impact of "quantum objects" on measuring instruments or, more generally, on the physical world to which we can relate directly through observation and measurement. As I explained in Chapter 1, Heisenberg's original (matrix) formulation of quantum mechanics conformed to this view more readily, and may be seen as shaping Bohr's view all along. His new kinematical elements, matrices, and the formal equations (retained from classical physics but now applied to these matrices) in fact relate, by means of statistical predictions, to the observable radiation effects, rather than to the motion of electrons.

Accordingly, the argument offered in this study gives a new and more radical meaning to Bohr's argument concerning the irreducible discrimination "between the *objects* under investigation and the *measuring instruments* which serve to define, in classical terms, the conditions under which the phenomena appear" (*PWNB* 2, p. 50). As just explained, this discrimination is consistent with the indivisibility of phenomena themselves. Indeed Bohr sees "this necessity of discriminating in each experimental arrangement between [them]" as "a *principle distinction between classical and quantum-mechanical description of physical phenomena*" (*PWNB* 4, p. 81). For, while "in classical physics the distinction between object and measuring agencies does not entail any difference in the character of the description of the phenomena concerned," in quantum physics it does (*PWNB* 2, p. 50; also *PWNB* 3, p. 3).

This and related statements by Bohr may suggest (and have suggested to some) that while parts of measuring instruments are described by means of classical physics, the behavior (in space and time) of quantum objects is described by means of quantum-mechanical formalism. Bohr obviously says the former, but he clearly does not ever say and does not mean the latter. Instead, the difference between the two descriptions is that classical theory describes the classical world, specifically certain parts of the measuring instruments, while quantum theory relates, in generally in statistical terms, to the interaction between the measuring instruments, described classically (while interacting quantumly with quantum objects) and quantum objects, indescribable by means of either classical or quantum theory. In this sense of the predictive capacity of the quantum-mechanical formalism, the totality of possible individual quantum phenomena (or effects) and specifically the complementary character of some of them can be comprehensively and consistently accounted for by quantum theory, while it cannot be unaccounted for by means of the classical theories. The two types of formalism themselves remain ultimately incompatible, even though they must coexist in Bohr's interpretation.

As Bohr points out, "it is true that the place within each measuring procedure where this discrimination is made is in both cases largely a matter of convenience" (*PWNB* 4, p. 81). The arbitrariness of this place, however, or the arbitrariness of the "cut," as it is also called, becomes a logical and helpful feature of this interpretation, part of Bohr's solution of the problem of quantum measurement, rather than a problem, as it would be if one assumed that the quantum-mechanical formalism describes the behavior of quantum objects themselves. (As discussed earlier, this assumption is often made in discussions of the Schrödinger cat paradox and decoherence.) At this point of his reply to EPR, Bohr brings into consideration the so-called transformation theorems of quantum mechanics. The latter mathematically explain why the EPR experiment and the EPR predictions are possible, and, according to Bohr, "perhaps more than any other feature of formalism contribute to secure its mathematical completeness and its correspondence with classical mechanics" (*PWNB* 4, p.74, n.). For "by securing its proper correspondence with the classical theory the [transformation] theorems exclude in particular any imaginable inconsistency in the quantum-mechanical description, connected with a change of the place where the discrimination is made between object and measuring agencies. In fact it is an obvious consequence of [Bohr's] argumentation that in each experimental arrangement and measuring procedure we have only a free choice of this place within a region where the quantum-mechanical description of the process concerned is effectively equivalent with the classical description" (*PWNB* 4, p. 82).

What makes this last point (which also conveys the deeper epistemological aspects of Bohr's correspondence principle, at least as the latter functions in Bohr's later works) so crucial is that quantum objects are always on the other side of the "cut" and indeed may be rigorously defined accordingly. This point is, again, intimated by Heisenberg's original presentation of his matrix version of quantum mechanics, and discussed by Heisenberg himself in his own (unpublished) response to EPR, reproduced in Pauli's correspondence (Pauli 1979-1999, v. 2). In Bohr's interpretation, the predictions of quantum mechanics refer strictly to classically manifest effects upon the measuring instruments, which can be rigorously correlated with specific configurations involving quantum objects. It is only this correlation that enables any possible identification of such a measurement with the position or momentum of a quantum object. The correlation is, according to Bohr, real, or rigorous, while the identification can only be, in his words, "symbolic." Bohr, as we have seen, does say that "to measure the position of [a] particle can mean nothing else than to establish a correlation between its *behavior* and some instrument rigidly fixed to the support which defines the space frame of reference [or some equivalent classical-like configuration]" (*PWNB* 4, p. 79.) He is, however, careful to say "behavior," which is quite different from saying "position." The sentence also compels one to argue that in Bohr's scheme this "behavior" itself, too, cannot be described but only correlated with the behavior of the measuring instruments, the behavior describable by means of classical physics.

At one end, by virtue of their classical nature, the individual effects in question can be isolated materially and phenomenologically—we can perceive and analyze them as such—once an experiment is performed, although they cannot be isolated from their efficacy by even conceiving, let alone analyzing, their physical emergence. By contrast, at another end, quantum objects and processes can never be isolated, either materially (from the measurement process and measuring instruments) or perceptually or even conceptually (we cannot in principle conceive of what actually happens at that level or how it happens).

Reciprocally, however, Bohr's argument under discussion bears crucially on the question of the classical description of the measuring instruments involved. It follows that, rather than to the constitution of measuring instruments as a whole (as is often assumed by Bohr's interpreters), this description applies only to the effect-constituting parts of these instruments. Their, as it were, efficacious components are quantum and, as such, are responsible for the effects in question and the classical behavior or those parts that, in their quantum interactions with quantum objects, produce these effects. Naturally, these components and these interactions are subject to the quantum-mechanical treatment, including the uncertainty relations, and Bohr's epistemology.

In this respect, Bohr's interpretation may not be as different from von Neumann's and related interpretations, as it sometimes argued in view of the not suitably qualified (along the lines just indicated) claim that in Bohr's interpretation the apparatus is described classically. Von Neumann himself was quite aware of this proximity, as his view of the "cut" would appear to indicate.[32] It is true that in von Neumann's interpretation and other interpretations of that type the apparatus is treated as the quantum-mechanical, rather than classical, system.[33] However, on the one hand, as just explained, so is the interactive-efficacious part of the apparatus in Bohr. On the other hand, the motion of the "pointer" registering the measurement result, crucial to von Neumann's scheme, is subject to classical motion or, epistemologically, perception, and as such is in fact analogous, even if not identical, to the classically described parts of the apparatus in Bohr. In other words, von Neumann's scheme, too, has a classical component at the level of the ultimate, perceptible or "phenomenal" (both in the usual and in Bohr's sense) effects of measurement. There are, of course, other differences between both interpretations, in particular, again, as concerns whether the quantum-mechanical formalism in fact describes the behavior of quantum objects (including the quantum component of measuring instruments). My main point at the moment are qualifications, crucial but often disregarded, pertaining to Bohr's argument concerning the classical behavior of measuring instruments and its significance for quantum measurements.

Thus, as I stress in this study, the measuring apparatus in Bohr is not simply classical, as is nearly universally claimed. It is both classical and quantum. Rigorously

---

[32] See the final chapter, "The Measuring Process," of his *Mathematical Foundations of Quantum Mechanics* (von Neumann 1983, pp. 417-445), which refers to Bohr (p. 420).

[33] See (Mittelstaedt 1997) for arguably the most refined version of this interpretation, seen as "the minimal interpretation" and for an elegant and illuminating discussion of its key features by Mittelstaedt, which, in my view, again, brings his interpretation closer to Bohr than Mittelstaedt himself appears to see it.

considered, the situation has far reaching implications for both quantum epistemology and the specific character of the mathematical formalism of quantum mechanics (in particular, its dependence on complex rather than real mathematical entities), although the latter subject exceeds my, primarily epistemological, scope here. It is worth keeping in mind, however, that Bohr's claim that "in each experimental arrangement and measuring procedure we have only a free choice of this place within a region where the quantum-mechanical description of the process concerned is effectively equivalent with the classical description" exceeds the limits of his interpretation and appears to be uncircumventable in general.

In sum, in Bohr's scheme the classical description applies only to physical effects manifest in certain parts of measuring instruments, which at the same time have an equally irreducible quantum component and (just as in von Neumann's scheme) interact quantum-mechanically with quantum objects. It is, as I have argued throughout, this interaction that is responsible, is the efficacity, of the classical effects in question.

It may appear that some interpretations of quantum mechanics allow us at least to directly *refer* to, if not quite to describe, quantum processes themselves and their key features, while bypassing measurement as an (irreducibly) constitutive component of quantum physics. Measurement is, now, seen as auxiliary, as in classical physics. Hence, such interpretations may be seen, and are sometimes offered, as realist, and, accordingly, as epistemologically different from that of Bohr. This would allow one to argue to have at least competing and, to many, epistemologically more palatable alternatives to, if not counter-arguments against, Bohr's interpretation. I am, let me qualify, not referring to the hidden variables theories and related nonlocal theories (these are mathematically different from quantum mechanics) or the many-worlds interpretation of quantum mechanics. These theories do not retain the standard features of quantum mechanics under discussion (and Bohmian theories employ a different formalism). I have in mind (in addition to some earlier interpretations , such that of Feynman) certain versions of the modal interpretation (specifically that of Bas van Fraassen), of the consistent histories interpretations (specifically those of Robert Griffiths and Roland Omnés, which may be further linked to decoherence). These are, arguably, among the closest to Bohr's among such interpretation, except for Bohr's epistemology (van Fraassen 1991, Griffiths 2003, Omnés 1994, Omnés 1999). As I have indicated, von Neumann's and related interpretations, such as Mittelstaedt's minimal interpretation are more grounded in the concept of measurement (Mittelstaedt 1997). Accordingly, I shall put them aside for the moment. Both van Fraassen and Omnés associate their interpretation with the Copenhagen interpretation; and Mermin's indicates further proximities between Bohr's and the histories interpretations of Griffith and Omnés (Mermin 1998a, p. 763).[34] I shall

[34] While also aiming to avoid the constitutive role of measurement in quantum mechanics, Mermin's own (Ithaca) interpretation (Mermin 1998a), offers a more subtle

refrain from the *definitive* assessment concerning such claims and offer instead the following set of observations, indicating what a Bohrian response concerning the possibility of this type of claims would be.

Bohr, I argue here, would deny that the quantum-mechanical formalism can unambiguously refer to quantum objects and processes in terms of space-time concepts in the way it can be done in classical physics or even in relativity, which already introduces significant complications in this respect (*PWNB* 2, pp. 40-41). As we have seen, he noted that "even in the indeterminacy relations, we are dealing with the formalism which defies unambiguous expression in words suited to describe classical physical [space-time] pictures" (*PWNB* 2, p. 40). Ultimately this formalism defies all "unambiguous use of space-time concepts," which, as discussed in detail in this chapter, are instead "confined to recording of observations which refer to marks on a photographic plate or similar practically irreversible amplification effects" (*PWNB* 2, p. 51). If one is concerned with the physical meaning of a theory as referring to the processes occurring in space and time, in Bohr's interpretation, what takes place in space and time is a particular configuration of phenomenal effects. Bohr saw the mathematical formalism of quantum mechanics as correlative to this situation, including as applied to the EPR experiment.

As discussed in Chapter 2, this formalism is highly symbolic and non-visualizable in its very nature, especially by virtue of using complex numbers, which could be related to observations, always recorded in real numbers, by means of ad hoc rules, such as Born's square moduli rule, projection postulate, and so forth. What would it mean to speak of the physical reality, of the physical meaning (in space and time) of such objects themselves, or of structures such as a linear superposition of vectors? Dirac, who may be seen as one of the originators of the approaches under discussion and the view that the quantum-mechanical formalism pertains to quantum object themselves, clearly denies that linear superposition can have any physical meaning. Accordingly, his epistemology, too, may be much closer (which is not to say identical) to Bohr's than it may appear, quite similarly to the case of von Neumann's views, as discussed above (Dirac 1995, pp.10-14).[35] In other words, such mathematical objects cannot realistically represent or directly refer to space-time processes (even noncausal ones); they may not

---

case, since it does not attributes physical properties (the "correlata" of quantum correlations) to quantum objects, or indeed to anything. It sees only quantum correlations themselves (without correlata) as having objective reality that physics can speak of. The subject would require a separate discussion, however, which I offer elsewhere (Plotnitsky 2003).

[35] It is not always easy to ascertain the degree of this proximity in the case of Dirac or others close to Bohr, such as Heisenberg and Pauli, since their epistemological views are not explicated to the same degree as Bohr's. By contrast the differences between Bohr's views and Einstein's (or those closer to Einstein, such as Schrödinger's or Bell's) are immediately manifest.

even constitute a mathematical, at least geometrical, phenomena in the sense of allowing us to conceive of them spatially.[36] Indeed, the practice (enabled by Dirac's and then von Neumann's versions of quantum-mechanical formalism) of using the Hilbert-space formalism and bra/ket-vector notation in explaining "what is actually going on" may be seen as a manifestation of this impossibility. This may also change our understanding of what it means to offer a description of a physical process, as (differently) Mermin and Haroche appear to suggest (Mermin 1998a, Haroche 2001). Such a change would, however, be consistent with Bohr's view insofar as one maintains that in speaking in terms of space-time behavior and classical-like physical variables or concepts we can only refer to the effects of the (quantum) interactions between (indescribable) quantum objects (and classically described) parts of measuring instruments.

As indicated earlier, spin (whose ket-vector description in a two-dimensional Hilbert space is particularly elegant and simple) offers an especially illustrative case and a partial model for a general understanding of the situation (and it may have served Bohr as both in shaping his thinking on quantum epistemology). Spin has no classical counterpart and we can only access it through its effects upon measuring devices and speak of it only by way of loose metaphorical analogies, which inevitably break down at a certain point. Now, insofar as we can refer to, say, the position and momentum of a particle, the situation is in fact quite similar epistemologically. It is different insofar as the *classically described* parts of measuring instruments cannot be assigned spin, while they have of course momentum or position (which may also serve as a reference frame, as, say, in the double-slit experiment). (Hence, as I said, the analogy with spin is only partial.) More crucially, in Bohr's interpretation, the situation is analogous insofar as any "position" or "momentum" impact of the interactions between quantum object and measuring instruments upon the latter does not, in all rigor, measure a classical-like variable of "position" or "momentum" of the quantum object involved. Once again, due to complementarity, we cannot measure both variables even as pertaining to measuring instruments, which variables are, thus, themselves subject to the uncertainty relations (*PWNB* 4, p. 77). In this sense, it would not be quite accurate to speak, as Bohr does in his reply to EPR, of the momentum exchange between a particle and the apparatus, unless provisionally, as Bohr appears to be doing in order counterargue EPR's argument, defined, via their concept of physical reality, in terms of properties of quantum objects (*PWNB* 4, p. 76). Bohr subsequently modified his argument in any event, along with his interpretation of all quantum phenomena, in terms of his concept of phenomenon, not yet in place in his reply to EPR.

---

[36] As noted in Chapter 2, these considerations are significant even in the case of spin, where we deal with finite-dimensional Hilbert spaces, but are more immediately manifest in the case of standard kinematical and dynamical variables and, hence, infinite-dimensional Hilbert spaces.

In sum, first, the standard quantum-mechanical formalism in any of its form does not offer a description of the behavior of quantum objects in space and time, but only a possible link to such a behavior (the nature of this link being subject to interpretation). This argument is not so surprising given the mathematical character of the standard quantum-mechanical formalism and (I say this with caution) *may*, accordingly, pertain to all interpretations of this formalism. Secondly, more radically, in Bohr's interpretation, the quantum-mechanical formalism cannot refer or even relate in any way, however indirectly, to the independent behavior of quantum objects but only to their interactive impact upon the measuring instruments involved. Naturally, we can still use such references for the sake of convenience and economy of discourse, while keeping their provisional (analogical, symbolic, metaphorical, and so forth) status in mind.

Accordingly, Bohr's concept phenomenon may be more than merely a possible conception of physical, space-time, situations compatible with the quantum-mechanical formalism, using the observational data in question, and thus only a part of Bohr's interpretation. Instead, this concept may be more fundamentally correlative to the quantum-mechanical formalism, insofar as anything in space and time rigorously accountable by its means is possible, and thus exceeds the limits of Bohr's interpretation. By contrast, nothing in regard to quantum processes may be ascertained. That, however, would appear to imply (as I said, I refrain from a definitive claim here) the following conclusion as concerns the interpretations in question at the moment, given that the mathematical formalism involved is that of standard quantum mechanics (certain nuances and rules of applications are introduced, especially in histories interpretations). Once the arguments of these interpretations are related to anything that can actually occur in space-time, Bohr's phenomena and, hence, measurement may have to be reinstated as constitutive rather than merely auxiliary. For example, arguments, such Omnés's, that the macro-world is ultimately subject to quantum-mechanical, rather than classical, description may be retained, but may need to be adjusted insofar as we can consider the aspects of this description only at the level of effects (Omnés 1994, Omnés 1999).[37]

This might make classical physics indispensable, even though ultimately inadequate, and may make it impossible to apply quantum mechanics to objects, such as the universe as a whole, which we cannot, in principle, consider as objects under observation and measurement. But then it is far from self-evident that such concepts as object, or wholeness and the universe, can apply to "the universe" as a "whole," even leaving aside quantum-epistemological complications of applying all such terms, as here considered.

---

[37] See, however, (Leggett 1988) and (Garg 2001), for one of the most significant programs, initiated by Anthony J. Leggett, along these lines of questioning quantum mechanics. These issues are subjects of continuing research. Cf., for example (Friedman, at al 2000).

3. EPR'S ARGUMENT AND BOHR'S RESPONSE

EPR begin their article by proposing the following, apparently natural, criterion of physical reality: "If, *without in any way disturbing a system*, we can predict with certainty the value of a physical quantity, then there exists an element of physical reality corresponding to this physical quantity" (EPR 1935, p.138; emphasis added). The argument itself is ingenious and, as Bohr noted, "remarkable for its lucidity and apparently incontestable character" (*PWNB* 2, p. 59).

"Apparently!" For, it may indeed appear that EPR's criterion of physical reality, just stated, applies in quantum mechanics as well, *without any further qualification*, just as it does in classical mechanics, as EPR claim. As they say: "Regarded not as a necessary, but merely as a sufficient, condition of reality, this criterion is in agreement with classical as well quantum-mechanical *ideas* of reality" (EPR 1935, p. 139; emphasis added). The reasons for this *apparent* applicability or, again, apparent unqualified applicability of EPR's idea of reality to the phenomena in question are as follows. In view of the uncertainty relations, it is only a joint *simultaneous* measurement, determination, or prediction (these must be carefully distinguished in turn) of two variables involved in the quantum-mechanical physical description that is impossible. A measurement, determination, and prediction of the value of a single variable are, in principle, possible with any degree of precision (within the capacity of our measuring instruments, while uncertainty relations are not affected by this capacity, but only by the value of Planck's quantum of action, h). Such a prediction is also possible in the EPR experiment, even though it deals with physical variables *associated with* two spatially separated quantum objects, (A) and (B), that have previously been in interaction, and it allows one to predict the values of certain physical quantities associated with one of these objects by making measurements upon the other.[38] Quantum mechanics allows one to

---

[38] As noted earlier, there is always a nonzero probability that our attempts to verify such predictions (or our attempts to make the measurements necessary to make these predictions) will not encounter the objects in question, for example, the second object of an EPR pair (or the first object after the initial preparation of the pair). It is not that the formulas for the wave function used by EPR do not give us the probability equal to unity for the predictions in question. They evidently do, which fact is stressed by Bohr from the outset of his reply and is crucial for the present discussion as well. However, the specific features of EPR's thought experiment, including the wave function they consider and the possible realization of it as an actual experiment, involve certain idealizations that allow us to disregard these factors. These idealizations were not missed by Bohr (*PWNB* 4, p. 79, n.). I shall bypass *some* of these complications here, since they do not affect the present argument, but only some of them, since this argument

make predictions concerning one or the other conjugate variables involved (although not both simultaneously) associated with one of these objects, (B), on the basis of a measurement on the corresponding variable associated with the other, (A). Thus, by using quantum mechanics, one can predict an outcome of a future position or, conversely, momentum measurement associated with the object (B) on the basis of the outcome of, respectively, a position or a momentum measurement performed on the other object, (A). The situation may be (and, with Bell, was) redefined in terms of correlated events of measurement for "identically" prepared EPR pairs. I place "identically" in quotation marks because we can only control the identical preparation of the measuring instruments involved, given that this behavior is classical, that is, the preparation of those parts of such instruments that are relevant to measurement, since measuring instruments also contain quantum strata, which enable their interaction with quantum objects. Unlike, however, in the phenomena considered in classical physics, one can never control the outcomes of thus repeated events. In other words, in general, identically prepared experiments lead to different recordings, including, importantly and even crucially, in the EPR case. This fact implies that the relationships between any proper formalism and observations must be statistical, in other words, the same as they are in quantum mechanics. Repeating the same experimental set-ups allows one to verify the fact of such correlations experimentally without an appeal to quantum mechanics, which, however, is in accord with these data, that is, it predicts these outcomes correctly.

Once the two objects in question are separated after their interaction (which "entangles" them, or rather the Hilbert-space state vectors enabling our possible predictions concerning them), quantum mechanics allows us to simultaneously predict both the *distance between* the two objects and *the sum of their momenta*. This is possible, given that the corresponding Hilbert-space operators commute, even though *quantum mechanics* prevents us from establishing or predicting both the position and momentum of either object simultaneously, in view of the uncertainty relations. Or, as Bohr argues, it is nature that prevents us from doing so, while quantum mechanics is a reflection of what nature allows or does not allow us to do, which is a crucial difference, missed, Bohr also argues, by EPR, but one of the focal points of Bohr's counterargument. With the two objects or, again, more accurately, our possible predictions concerning them thus entangled and with these additional quantities in hand, by *measuring* (which can be done with any degree of precision, at least in principle or in the idealized situation) either the position or, conversely, the momentum of the one object (A), one can *predict exactly* either the position or the momentum, respectively, for the second object (B), without interfering with it in any way. This is very easily shown, in fact, as Bohr notes, even more generally than by EPR, via the transformations

---

essentially depends, as does that of Bohr, on the irreducibly statistical nature of the quantum-mechanical predictions.

theorems, mentioned earlier (*PWNB* 4, pp. 74-75, n.; pp. 79-80). Such quantities become "correlated," once one measures this predicted quantity and repeats the experiment (with "identical" preparations of the pairs) many times.

Bohr agrees that quantum mechanics allows for and enables such ("at-a-distance") predictions concerning the EPR correlations and entanglement, along all other quantum phenomena, in either sense, since the EPR-type experiment, whether in its original terms or in terms of spin, can be adjusted so as to refer only to the outcomes of measurements rather than to quantum objects.[39] Indeed, as Bohr observes, in a certain sense, the EPR conditions apply to all quantum-mechanical predictions. In any quantum measurement, we can predict either the position or the momentum of a particle after a preceding measurement took place and on the basis of this measurement, and hence without interfering with the object and without assuming that we can define either quantity independently of measurement (*PWNB* 2, p. 57). In fact, the particle and those parts of the measuring instruments involved that interacted with it become entangled.[40] Hence, at the time of determination in question, there is no physical interaction either between the two particles or between any measuring apparatus and the second particle. In the EPR situation, which involves two particles, we have a slightly more complicated, but not fundamentally different, case, which also allows us to make prediction with the probability equal to unity, as against to the standard case, where such probabilities are, in general, not unity.

Bohr does not agree, however, with EPR's conclusions. In Bohr's summary, "According to their criterion, the authors [EPR] therefore want to ascribe an element of reality [associated with the same quantum object (B)] to each of the quantities represented by such variables. Since, moreover, it is a well-known feature of the present formalism of quantum mechanics that it is never possible, in the description of the state of a mechanical system, to attach definite values to both of two canonically conjugate variables, they consequently deem this formalism to be incomplete, and express the belief that a more satisfactory theory can be developed" (*PWNB* 4, p. 74). In other words, Bohr does not see quantum mechanics as incomplete, since he does not think that it follows that the possibility of predicting with the probability unity of either of the two quantities in question allows one to assign both to the same quantum object. In fact, as I argue here, he does not think, at least in his ultimate view of the situation, that even one such a quantities could ever be assigned to a quantum object (as opposed to a certain part of the measuring apparatus involved), either on the basis of a prediction or even on the basis of a measurement. Hence, there is no contradiction of the kind EPR contemplate

[39] Given the main concerns of this study, I follow Bohr in presenting the subject in terms of the double-slit experiment, as explained in Chapter 1, and the position-momentum complementarity, rather than in terms of spin, as is more common in current discussions and, in some respects, more effective.

[40] Cf. S. Haroche's argument (Haroche 2001).

between what quantum mechanics does and what is in principle possible, what nature allow us to do. At least, EPR's argument does not demonstrate such a contradiction.

At the same time, however, the situation does not, in Bohr's interpretation, imply nonlocality, as some suggest, which also allows one to circumvent the alternative between locality and completeness of quantum mechanics, which is suggested by EPR and becomes the main focus of Einstein's subsequent arguments concerning the EPR type experiments. In other words, entangled quantum objects or (Bohr, though, shuns this language) "states" exist, but this only means a) that particular forms of experimental detection (phenomena or effects) are possible, and b) that the quantum-mechanical formalism allows us to predict such effects. Accordingly, nonlocality, which is entailed neither by a) nor by b), need not follow, unless of course it is independently derived from the formalism itself, which does not appear to be the case thus far. These facts, combined, thus also allow one to see quantum mechanics as both a complete and local, even if not necessarily unique, theory of the phenomena in question in it.

It has also been argued, including, as noted earlier, by Einstein, that the question of locality is not relevant for Bohr, or that he allows for the nonlocality of quantum mechanics or quantum phenomena. It would be difficult to agree with either view. Bohr's concern with the relationships between quantum mechanics and relativity is manifest in his reply to EPR and most of his subsequent (and preceding) writings on the subject. He states in a key (but often ignored) note at the end of his reply: "the relativistic invariance of the uncertainty relations ensures the compatibility between the argument outlined in the present article [i.e., his reply to EPR] and *all exigencies of relativity theory*" (*PWNB* 4, p. 82, n.).[41] Indeed, as the argument (mentioned as forthcoming in the same note) of "Discussion with Einstein," advanced under the impetus of Einstein's criticism, makes clear the proper application of quantum-mechanical formalism and specifically of the uncertainty relations in fact depends on general relativity. The discussion on the subject between Einstein and Bohr took place well before the EPR article, in 1930, and, as Bohr observed, it contained all the key epistemological elements of the EPR argument (*PWNB* 2, p. 57).

For Bohr, relativity and hence, locality was a given, an "axiom" or a "law of nature," confirmed by experiment, just as the uncertainty relations were a given, "a law of nature," confirmed by experiment and adequately accounted for and indeed a consequence of quantum theory, or as the classical description of certain parts of measuring instruments were a given, corresponding to the fact that on the macro-level classical physics still holds. The latter view further reinforced by the "correspondence" argument, considered earlier, that the "cut" between classical and quantum-mechanical description could only be made in the region where both coincide. The initial challenge answered by

[41] See the version of Bohr's reply published in *QTM* for a full text of this note, abbreviated in the version published in *PWNB* 4 (*QTM*, p. 150, n.).

complementarity, to some degree even as it was developed in the Como lecture, was to find an interpretation of the quantum-mechanical phenomena and formalism, especially as manifest by the wave-particle complementarity and the uncertainty relations. Einstein's subsequent arguments, especially those of the EPR type, offered a new challenge, that of proving that complementarity is, or may be refined to be, both complete, within its scope, and consistent with relativity. In answering this challenge, Bohr's ultimate version of complementarity allows him to see quantum theory and data as fully consistent with all three (within their proper scopes), quantum mechanics, relativity, and classical physics, and brings all three together, even if at the cost (again, exorbitant to Einstein and, following him, many others) of Bohr's radical epistemology. The latter deprives us of the possibility of knowing how quantum entanglement and correlations are ultimately possible in physical (space-time) terms. In this view (and with preceding analysis in mind), quantum mechanics only predicts certain effects of the ultimate constituents of matter upon their quantum interaction with measuring instruments, but does not describe either quantum objects or these effects. These effects are subject to a description, but *not prediction*, in terms of classical physics, along with the behavior of those parts of measuring instruments that involve these effects (such as a change in momentum of a certain part of a given apparatus).

Bohr argues that "the apparent contradiction in fact discloses only the essential inadequacy of the customary viewpoint of natural philosophy for a rational account of physical phenomena with which we are concerned in quantum mechanics." Instead, in accordance with the preceding discussion, the irreducibility of this interaction "entails the necessity of the final renunciation of the classical ideal of causality and a radical revision of our attitude towards the problem of physical reality"(*PWNB* 4, p. 74).[42] The contradiction in question is, as noted above, that between what quantum mechanics does and what nature allow us to do. Bohr's statement may be too strong, even though it may one day prove to be right. At the very least, however, this view makes it possible to develop a local interpretation of quantum mechanics, as well as, as a kind of philosophical bonus, a new form of epistemology.

Accordingly, quantum entanglement and the question of locality make "interpretation" even more significant in quantum mechanics. Short of a particular, indeed very specific, interpretation, it may not be possible to counter Einstein's arguments. Even if at an epistemological cost (acceptable to Bohr, exorbitant to Einstein), Bohr's interpretation ensures the compatibility of quantum mechanics with relativity, the desideratum equally defining his and Einstein's view. Is Bohr's

---

[42] I omit Bohr's intermediate propositions concerning the interactions between quantum objects and measuring instruments. While they are important in explaining the reasons for Bohr's argument, they are fully consistent with the preceding analysis, presented in terms of "Discussion with Einstein."

interpretation the only such interpretation currently available? Maybe not, although, as I have suggested, it is at least useful (if not necessary) in helping other interpretations to make this compatibility more rigorous. Are more radically different interpretations of quantum theory possible, say, those that Einstein would accept as "more complete conceptions"? That is, is the epistemology in question inevitable given quantum theory or even already quantum data or only consistent with it? It is difficult to say, and it is the latter, quite enough in itself, that is my argument here. Bohr, too, especially in print, appears to argue primarily for the latter, although he did not appear to think (rightly) that there were other fully developed such interpretations at the time of his writing. This view of Bohr's argument is not inconsistent with his claims to the effect that "in quantum mechanics we are not dealing with an arbitrary renunciation of a more detailed analysis of atomic phenomena, but with a recognition that such an analysis is, *in principle,* excluded." As is clear from Bohr's accompanying elaborations, this claim may be (and may need to be) read as applicable only within Bohr's interpretation, seen by Bohr as enabling one to view quantum mechanics as a fully consistent and, within its proper scope, complete physical theory.

These are these considerations that enable Bohr to argue that EPR's criterion of reality contains an essential ambiguity when it is applied to quantum phenomena. I would now to explain this argument by offering a reading of Bohr's famous passage on "the essential ambiguity" from his reply to EPR. Bohr writes:

> From our point of view we now see that the wording of the criterion of physical reality proposed by Einstein, Podolsky and Rosen contains an ambiguity as regards the meaning of the expression "without in any way disturbing a system." Of course there is in a case like that just considered no question of a mechanical disturbance of the system under investigation during the last critical stage of the measuring procedure. But even at this stage there is essentially the question of *an influence on the very conditions which define the possible types of predictions, regarding the future behavior of the system.* Since these conditions constitute an inherent element of the description of any phenomenon to which the term "physical reality" can be properly attached, we see that the argument of the mentioned authors does not justify their conclusion that quantum-mechanical description is essentially incomplete. On the contrary this description, as appears from the preceding discussion, may be characterized as a rational utilization of all possibilities of unambiguous interpretation of measurements, compatible with the finite and uncontrollable interaction between the object and the measuring instruments in the field of quantum theory. In fact, it is only the mutual exclusion of any two experimental procedures, permitting the unambiguous definition of complementary physical quantities, which provides room for new physical laws, the coexistence of which might at first

> sight appear to be irreconcilable with the basic principles of science. It is just this entirely new situation as regards the description of physical phenomena that the notion of *complementarity* aims at characterizing. (*PWNB* 4, p. 80; Bohr's emphasis)

This is the most famous and most cited elaboration of Bohr's reply and perhaps of all of Bohr's writings. It was seen as crucial by Bohr himself, who cited and commented on it in "Discussion with Einstein," which tells us directly what is the nature of the ambiguity of the EPR criterion of reality and their argument. Bohr defines his argumentation in his reply as aiming at "bringing out *the essential ambiguity involved in reference to physical attributes of objects* when dealing with phenomena where no sharp separation can be made between the objects themselves and their interaction with measuring instruments" (*PWNB* 2, p. 61). The ambiguity of the EPR criteria is this ambiguity. This message is rarely heeded by most commentators on Bohr's argument, who try to understand this elaboration on the basis of experimental facts and general theoretical arguments of quantum mechanics rather than in Bohr's own terms. It took Bohr a while to work out the ultimate version of complementarity and, with it, of his interpretation of the EPR experiment, as considered here, which, as he notes on the same occasions, helps him to mitigate "the inefficiency of expression" of the passage from his reply in question, although Bohr appears to refute EPR's argument on other grounds as well. This "inefficiency of expression," he says, "must have made it very difficult to appreciate the trend of the argumentation aiming to bring out the essential ambiguity involved in a reference to physics attributes of objects when dealing with phenomena where no sharp separation can be made between behavior of the objects themselves and their interaction with the measuring instruments" (*PWNB* 2, p. 61).

The invocation of "ambiguity" in and Bohr's claim that this ambiguity arises "as regards the meaning of the expression 'without in any way disturbing a system'" is sometimes taken (including by Einstein) to mean that there is in fact, contrary to the EPR argument, some form of "disturbance" or "influence" with respect to the spatially separate system or quantum object—the second object (B) of the EPR pair.[43] It is difficult to argue that this is what Bohr has in mind here. There does not appear to be anything to support such a view either in the elaboration in question or in Bohr's preceding argument in his reply. The latter is important, as the opening phrase, "from *our point of view* we now see," indicates, in particular in stating that Bohr has a (local) interpretation within which the EPR criterion is ambiguous, but it, too, is often ignored

[43] Among such commentators are B. d'Espagnat (d'Espagnat 1989, pp. 94-95, 255), J. Faye (Faye 1991, pp. 181-182), H. Folse (Folse 1987), J. Honner (Honner 1987, pp. 125-141), and H. Stapp (Stapp 1987, p. 162).

by critics.[44] In order to properly understand Bohr's meaning, one must consider other propositions in Bohr's reply and elsewhere on what can and cannot be unambiguously done in quantum mechanics. Once one does so, it becomes apparent that the "ambiguity" here need not be read as stating that, contrary to EPR, there is in fact a "disturbance" in making the EPR predictions, which, thus, invalidates their argument. Instead this ambiguity has to do with the conditions (including the necessary interpretation of the measurements involved) under which, in the EPR situation or in quantum mechanics in general, an unambiguous meaning can be assigned to terms and concepts, such as EPR's "elements of reality," that define the situation in question. These are the conditions of the interaction between quantum objects and measuring instruments, and ultimately of phenomena in Bohr's sense. Bohr does not have any "disturbance" in mind. As I said, he *agrees* with EPR that we cannot allow for any "disturbance," physical or other, of anything that is spatially separated from actual physical measurement.

To some degree, the reading of Bohr's elaboration turns on how one reads Bohr's "But" here. One way of reading this "But" (and this is what is most often done) is to proceed from "Of course there is in a case like that just considered no question of a mechanical disturbance of the system under investigation during the last critical stage of the measuring procedure" to "But even at this stage there is essentially the question of *an influence on the very conditions which define the possible types of predictions, regarding the future behavior of the system*," and disregard the next sentence: "Since these conditions constitute an inherent element of the description of any phenomenon [of quantum physics] to which the term 'physical reality' can be properly attached ['unambiguously applied'], we see that the argument of the mentioned authors does not justify their conclusion that quantum-mechanical description is essentially incomplete." What, however, can, and in my view should be, done is this. One starts with "Of course there is in a case like that just considered no question of a mechanical disturbance of the system under investigation during the last critical stage of the measuring procedure," thus, taking locality as an axiom, and proceed to reading Bohr's "But" as joining the next two sentences together (by, as it were, inserting an "and" connecting them): "But even at this stage there is essentially the question of *an influence on the very conditions which define the possible types of predictions, regarding the future behavior of the system* [and] since these conditions constitute an inherent element of the description of any phenomenon [of quantum physics] to which the term 'physical reality' can be properly attached ['unambiguously applied'], we see that the argument of the mentioned

[44] I am not offering a discussion of these elaborations here because the argumentation they contain largely amount to Bohr's argument as considered in this study. This discussion, it is true, adjust them, as Bohr himself did, to Bohr's ultimate view of the situation and of quantum mechanics in terms of his concept of phenomena. While it makes for an easier counterargument to EPR, this adjustment does not affect my analysis of the passage in question.

authors does not justify their conclusion that quantum-mechanical description is essentially incomplete."

If one follows this reading and takes the preceding discussion into account, it is clear that Bohr speaks of an ambiguity as concerns the expression "without in any way disturbing the system" not because there is a disturbance or influence of any kind, but because the whole reference to the system in question is "ambiguous" in Bohr's interpretation. This, then, is why Bohr says that the EPR criterion of physical reality contains "an [essential] ambiguity as regards the *meaning* of the expression 'without in any way disturbing the system'" (*PWNB* 4, p. 75). It concerns the question of an unambiguous definition of the system itself rather than indicates that the system is in fact disturbed or influenced and that the EPR application of their criteria in quantum mechanics is ambiguous for that reason. In fact, it is an obvious consequence of Bohr's interpretation that in any quantum-mechanical measurement the language of "disturbance" cannot apply in the sense that there are any undisturbed properties of quantum objects that are then disturbed by measurement (*PWNB* 2, pp. 63-64). Nor, however, does one need to conclude that, because we cannot speak of the properties of objects apart from when experiments establishing them are performed (in Bohr's view we cannot do so even then), the two quantum systems in question in fact form an inseparable nonlocal whole (as they would in Bohm's theory). Consistently with the formalism of quantum mechanics, these systems are seen as spatially separated or rather whatever can in fact be ascertained concerning them in space-time, that is, any phenomena (in Bohr's sense) associated with them, would be spatially separated.

It follows, then, that the "influence" in Bohr's statement does not refer to any influence, physical or not, upon the system associated with the second EPR particle (concerning which we make predictions but upon which we do not perform measurements). Instead it refers to the physical influence upon the (measuring) system associated with the first EPR particle (upon which we do perform measurements and with which we, hence, interfere) and on the conditions of such measurements as concerns the physical quantities involved. As a result, this influence of course—this is indeed Bohr's whole point—"influences" what kind of predictions we can or cannot make concerning the second particle. Hence it defines the conditions of the always irreducibly two mutually exclusive phenomena in the EPR-type experiments or (and these are, again, essentially the same conditions) the conditions of all quantum-mechanical predictions. If the apparatus is set up for measuring one complementary variable for the measuring system linked to the first particle, then within this setup the other variable cannot be *unambiguously defined*—we are absolutely precluded from doing so—for either particle of the EPR pair. Or more accurately, we cannot define it for the first system and hence cannot predict the corresponding variable for the second. For, "we have in each experimental arrangement suited for the study of proper quantum phenomena not merely to do with an ignorance of the value of certain physical quantities, but with *the impossibility of defining these quantities in an unambiguous way*" (*PWNB* 4, p. 78; emphasis added). Accordingly, as Bohr states already in his preliminary note in *Nature*, "the procedure of measurement has *an essential influence* on the conditions on which the very *definition* of the physical quantities in question rests" (Bohr 1935, *QTM*, p. 144; emphasis added). This influence is thus that upon the conditions of an unambiguous definition of the quantities in question. These conditions physically pertain to the system associated with the first particle, where direct physical measurements are performed, even when the prediction and the possibility of definition themselves concern the second particle (which, or anything that can be associated with it, is not physically influenced). This is why, while Bohr indeed says that there is "*an influence on the very conditions which define the possible types of prediction regarding the future behavior of the system*," he never says that anything disturbs, interferes, or even influences the second system. The former need not, and in Bohr's argument does not entail, the latter. Nor, as noted above, does he see any other (nonphysical) disturbance or influence upon the second particle or, again, any system (at-a-distance) locally associated with it. Ironically (vis-à-vis Einstein's expectation), any classical-like complete theory of the type Einstein wanted and thought possible by virtue of his arguments of the EPR type that would predict these data now appears to be nonlocal in view of Bell's theorem and related findings. The situation would be similar to Bohmian, hidden variables, quantum mechanics, where, however, nonlocality is an explicit consequence of the mathematical formalism (in any version of it available so far). Bell's theorem tells us that any classical-like (hidden variables) theory would be nonlocal, a finding further amplified by related theorems, such as the Kochen-Specker theorem, all of which, in the present view, appear to support Bohr's argument.[45]

We can now see that Bohr's "But" in the sentence referring to the influence in question need not be read as claiming that there is some physical influence, of whatever type, between two spatially separated systems, after Bohr's strong claim concerning the absence of any mechanical disturbance of the second particle, or whatever system is associated with it. Instead, if one takes into account Bohr's paragraph as a whole, this "But" indicates that Bohr questions the applicability of the EPR criterion of reality and

---

[45] I must bypass the debates concerning these theorems, specifically as regards how tight these arguments ultimately are. For a representative and still useful (although by now somewhat dated) selection, see (Cushing and McMullen 1989) and, for more recent commentaries (Ellis and Amati 2000). The view just expressed may be argued to be consistent with that of a significant majority of commentators. See, for example, N. D. Mermin's essays on the subject collected in (Mermin 1990) or his argument in (Mermin 1998a), (Gottfried 2000), (Bertlmann and Zeilinger 2002), and (Zeilinger et al. 2005), or, with a different valuation, most of J. Bell's articles on the subject in *Speakable and Unspeakable in Quantum Mechanics* (Bell 1987).

argumentation, while remaining in agreement with EPR's locality requirement and pointing towards an interpretation of quantum mechanics compatible with it.

Accordingly, given Bohr's interpretation and the very possibility of such an interpretation, it need not follow that quantum mechanics is incomplete by the EPR criteria (in this interpretation, inapplicable without ambiguity), which argument Einstein accepted to some degree. Nor does it follow that it is nonlocal, since, on the one hand, all the predictions enabled by the formalism are consistent with relativity. (The formalism itself is of course not relativistic corresponding to the scope of quantum mechanics, as against that of quantum electrodynamics.) On the other hand, all available arguments concerning nonlocality of (the standard) quantum mechanics, beginning with Einstein's, appear to make presuppositions which Bohr's interpretation does not make (which still appears to be the case), beginning and perhaps always amounting to making assumptions concerning the independent behavior of quantum objects. I would like to briefly comment in this context on the use of counterfactual logic in quantum mechanics.

Bohr's interpretation would inhibit the applicability to quantum theory of EPR's argument or related arguments by Einstein and most standard (perhaps all) counterfactual arguments, that is, arguments based on what could, but actually did not, happen, if certain (unperformed) experiments were "in fact" (a word perhaps inapplicable under these conditions) performed. Thus, in examining an outcome of a given double-slit experiment we can only argue from within one among the possible complementary (mutually exclusive) situations, defined by the particular experimental arrangement at hand, say, when both slits open and no counters installed. Our examination, however, cannot rely on (counterfactual) considerations of the other complementary situation, defined by the alternative arrangement, here the one that would allow us to determine through which slit particles pass. We cannot draw conclusions concerning the situation at hand by considering what could, but actually did not, happen to the quantum objects. In the EPR situation the circumstances of complementarity manifest themselves in the fact that, in practice, two runs of any EPR type of experiment, and hence two actual physical situations are always required in order to confirm the EPR alternative predictions. Just as in (and equivalently to) the double-slit experiment, there is no single physical situation of which these two situations would be parts or aspects, or from which they could be derived, of which they would be effects, since, as we have seen, the efficacity of the complementary effects involved is as individual, unique each time as are the effects themselves. Hence, the EPR and Einstein's other nonlocality arguments cannot avoid counterfactual logic.

Following Bell's theorem (which does not prove the nonlocality of quantum mechanics, but only of certain, epistemologically realist hidden variables theories, which allow for counterfactual reasoning), the assumption that counterfactual logic applies in quantum mechanics appears to be correlative to deriving nonlocality from quantum theory or quantum-mechanical data. These assumptions, however, may be seen as in turn supplementary rather than structural in quantum physics, as Bohr perhaps knew, and as both Arthur Fine and N. David Mermin emphasized (Fine 1989; Mermin 1998a). Indeed, most of such arguments rely, directly or implicitly, on attributing physical variables or quantum-mechanical descriptions to quantum objects rather than to the outcome of quantum measurement registered in measuring instruments, even though counterfactual statements themselves may refer to the latter. Or, again, they rely on the presupposition that both complementary situations may, at least in principle, be derived from a single common situation, know or unknown (or even unknowable). As I have just indicated, this presupposition is untenable in Bohr's interpretation and, it appears to follow (courtesy to Einstein), in any local interpretations of quantum mechanics. As Mermin observes, considered in terms of strictly statistical correlations, given a sufficient number of trials, quantum mechanics is local (Mermin 1990, pp. 108-109). Einstein to some degree realized this point, although he considered this way out of the dilemma "too cheap": "The interpretation of the ψ-function as relating to an ensemble also eliminates the paradox that a measurement carried out in *one* part of space determines the *kind* of expectation for a measurement carried out later in *another* part of space (coupling parts of systems far apart in space)"(Born 2005, pp. 205, 211). On the other hand, as Mermin shows, you cannot establish that quantum predictions, corresponding to these correlations, are strictly local at the level of individual events, which is of course what Einstein thought. However, as Mermin points out, the latter argument, just as Einstein's own, involves counterfactual reasoning (Mermin 1990, pp. 171-74).

Here, again, given the possibility of Bohr's interpretation, we face a simple "either/or" logical choice—of giving up either locality/relativity, a well-confirmed experimental feature of the physical world, or certain arbitrary, even if commonsensical and in terms of classical physics natural, assumptions, such as the possibility of applying counterfactual reasoning. It follows, for example, that Einstein's alternative between the "nonlocality and completeness" and the "locality and incompleteness" of quantum mechanics depends on his use of counterfactual logic, and thus offers us the possibility of ascribing to quantum mechanics "locality *and* completeness," while suspending the applicability of counterfactual reasoning. Bohr's interpretation of course does not depend on the latter. An intriguing possibility would be to consider giving up the assumption that certain parts of measuring instruments and, by implication, the macro-world are in fact described in terms of classical physics. The question would require a separate discussion, although it may be pointed out that the applicability of classical physics at this level is difficult to contest, and quantum physics has relied on it in other respects as well throughout its history, beginning with the correspondence principle. My main point at the moment is that Bohr's interpretation allows him to handle Einstein's argument, even if we adjust the latter argument to center on nonlocality and on the alternative of the nonlocality and incompleteness of quantum mechanics, as Einstein did later, rather than only on the incompleteness of it. It is worth keeping in mind, however, that Bohr's reply to EPR addresses the EPR argument as such, and it would be inaccurate to disregard or diminish this fact, or the character of the EPR initial argument in considering and assessing Bohr's reply, as is sometimes done.

Obviously, there remains the fact that we make predictions concerning the future behavior of the second particle as a result of interfering, substantially (by physically, mechanically engaging with measuring devices) and unsubstantially (by making meaningful statements, predictions, etc.), with the local measuring system associated with the first particle. That, however, is quite different from and indeed incompatible with any reading of Bohr's argument as implying nonlocality. The system associated with the first particle enables us to unambiguously define either one or another complementary variable (never both simultaneously) associated, via a corresponding measurement or phenomenon, with the second particle. A measuring system of the same type may be introduced for the second particle and, then, (once a sufficient number of "identically" prepared experiments is performed) correlated with the one associated with the first particle, once this second system is being in turn interfered with, for example, in order to measure the value of a given variable for the second particle. Such correlations are involved in Bell's theorem and its refinements and the discussions surrounding these findings and their implications, more recently around Greenberger-Horne-Zeilinger and Lucien Hardy types of experiments (refinements of the EPR type of experiments). Bohr's arguments obviously do not refer to these findings. On the other hand, what he says appears to be consistent with them or with the role of "correlations" in quantum mechanics more generally, a point that, as Mermin suggests, Bohr perhaps missed, or missed making in this form (Mermin 1998a, p.765, n.31). Nor, of course, could Bohr have had in mind more subtle features in subsequent developments around the EPR experiment(s), but his argument can, I think, be easily adjusted to these cases as well. My main point here is that quantum entanglement and locality are both maintained, and can be experimentally and theoretically understood within the epistemology of "effects," as here discussed.

How, then, is this possible, especially given the apparent impossibility of applying Reichenbach's common cause principle, according to which such correlations should be explained in terms of the preceding common history of the events involved? (Reichenbach 1956, p. 159) But then how do electrons "know," individually or collectively, that both slits are open to arrange themselves in the interference-like pattern? The answer of Bohr's interpretation is that we do not know how this ultimately is or can be (physically) possible, indeed we cannot know, or even conceive of it, since any further analysis that would, in principle, allow us to do so is, in principle, excluded. Bohr offers us the epistemology of knowable effects whose efficacity is ultimately inconceivable and, hence, the inaccessibility of "the quantum world" itself (to any knowledge or conceptualization that is or will ever be available to us, for example, as "quantum" or the "world," in any conceivable sense of either word, hence both potentially inapplicable) that allows one to rigorously maintain both the consistency and locality of quantum theory. Otherwise, however, this interpretation conforms to all standard requirements and desiderata of scientific research, and indeed enables us to advance where epistemologically classical accounts fall short. One can certainly regard most, indeed, it appears, so far all, of the advances in quantum theory in these terms, even though we often formulate them in more traditional terms. There is no special need to change this (and perhaps we can never quite do it), provided that we keep the provisional, rather than rigorous, nature of this classical-like terminology in mind. Bohr complementarity makes quantum mechanics prohibit doing or knowing in principle what *quantum mechanics itself*, as a physics theory, cannot do or know in practice. But quantum mechanics delivers plenty of what can be known, too, and different ways in which knowledge, including information in its technical sense, can be obtained and processed, for example, through quantum entanglement, which, among other things, enables quantum cryptography and computing. That this view entails the impossibility of any knowledge or certain conception concerning the ultimate constitution of nature has a positive role to play, since this impossibility enables knowledge and conceptions that might not be possible otherwise.

Chapter 4. Complementarity, Chance, and Probability

> [...] any comparison between quantum mechanics and ordinary statistical mechanics,—however useful it may be for the formal presentation of the theory,— is *essentially* irrelevant.
>
> —Niels Bohr

1. CHANCE AND PROBABILITY IN CLASSICAL AND QUANTUM PHYSICS

This chapter addresses the peculiar, arguably uniquely peculiar, character of chance and probability in quantum mechanics, at least in certain interpretations of it, such as Bohr's complementarity. Since, accordingly, the argument of this chapter depends on a particular (type of) interpretation of quantum mechanics and on the epistemology (fundamentally nonrealist in character) that this interpretation entails, this epistemology, as developed in this study, inevitably remains my equally primary subject here. Importantly, this particular view of chance and probability only pertains to the *physical* behavior of the systems considered and is defined by suspending causality even at the level of the individual entities involved, as against the view usually adopted in classical statistical physics, whereby the behavior of the individual entities comprising the multiplicities considered statistically is assumed to be causal. (As explained throughout this study, in Bohr's interpretation, these entities are not quantum objects themselves but instead certain individual experimental situations manifest in measuring instruments impacted by their interactions with quantum objects.) By contrast, the *mathematics* deployed for estimating the probabilities involved is not essentially different, which fact may indeed be seen as adding to the peculiarity of the situation that we encounter in quantum mechanics. I would like to begin by establishing the key terms of my discussion, including the most basic ones, such as "chance" and "probability," in part because their meanings vary in the discussions of quantum mechanics and their unspecified use may lead to confusion.

By "chance" I mean a manifestation of the unpredictable. A chance event is an unpredictable event, and, in general, it may not always be possible to estimate meaningfully whether a chance event would occur or with what probability, or sometimes to anticipate it as an event. On the other hand, probability considerations deal, theoretically or practically, with providing numerical estimates of occurrences of certain individual or collective events, specifically in accordance with mathematical probability theories, which theorize such estimations. Physically, however, such chance events, may or may not manifest some underlying causal dynamics unavailable to us, which defines, respectively, what I shall call *classical* and *nonclassical* chance and probability, such as that found in quantum mechanics in the present interpretation. By the expressions "fundamentally or irreducibly random" and "probabilistic" I refer to the chance and probability that are irreducible not only in practice (which may be the case in classical physics and is always the case in classical statistical physics), but also in principle, in other words, to the nonclassical chance and probability. The latter, I argue, defines quantum phenomena and quantum mechanics in the present interpretation. I choose the term "nonclassical," rather than quantum, first, because the concept itself is applicable beyond quantum physics, and, second, the chance found in quantum mechanics could be interpreted classically. This irreducibility of randomness in principle rather than only in practice, again, refers not to the *mathematics* of probability involved, but to the difference in the *physics* and, correlatively, *epistemology* found in quantum mechanics in the present interpretation vs. classical physics (in most interpretations of it) or certain alternative views of quantum phenomena and quantum mechanics.[46]

In particular, in classical physics, whatever the mathematics of probability might be, randomness and probability might be, and commonly are, seen as resulting from insufficient information concerning systems that are at bottom causal but whose mechanical complexity (usually defined by large numbers of their individual constituents) prevents us from accessing their causal behavior and making deterministic predictions concerning this behavior. As earlier in this study, I distinguish causality and determinism as follows. I use causality as an *ontological* category (in the sense to be further explained below) relating to the behavior of the systems whose evolution is defined by the fact that the state of a given system is determined at all points by their

---

[46]There might be arguments regarding what kind of probability theory (e.g., Bayesian, Kolmogorovian, contextual, and so forth) suits best quantum mechanics or quantum physics in general, including quantum statistical physics, which deals expressly with quantum multiplicities. See, for example, (Khrennikov 2004) on contextual probability and (Fuchs 2001) and (Fuchs 2003) on the Bayesian approach. There are also differences in how one actually calculates probabilities in classical vs. quantum physics, which differences are, as will be seen, also due to certain physical and epistemological factors and thus are relevant for my analysis here. These qualifications, however, do not fundamentally affect my argument concerning the lack of *essential* difference in the mathematics of probability used in classical and quantum physics (and besides we do employ different mathematical theories of probability in different cases in classical physics as well). Indeed this argument is supported the works just cited, however different the views expressed there may be from my own in other respects. Fuchs's argument is, however, epistemologically close to the one offered here as well, since the Bayesian approach is, arguably, closest to Bohr's view of quantum probability.

state at a particular point (usually at any given point). I use determinism as an *epistemological* category defined by our ability to predict the state of a system at any and all points once we *know* its state at a given point (again, usually such a knowledge at any point suffices). Thus, classical mechanics deals deterministically with causal systems; classical statistical physics deals with causal systems, but only statistically, rather than deterministically; and chaos theory deals with systems that are, in principle, causal, but whose behavior cannot be predicted even in statistical terms in view of their nonlinearity and, hence, their sensitivity to initial conditions. All these and other classical theories are *causal* insofar as they deal, deterministically or not, with systems that are assumed to behave causally.

These theories, at least in most versions or interpretations, are also, and in part correlatively, realist (keeping in mind that they deal with idealizations and models abstracted from actual objects in nature and their behavior). It does not automatically follow that deterministic theories are realist, since a possibly causal actual behavior of a system may not be mapped by our description of it, even though we can make reasonably good predictions concerning it. Classical mechanics or chaos theory is, however, realist insofar as such a mapping is assumed to take place, at least as an idealization or as a good approximation that such models provide. By contrast, classical statistical physics is not realist in the same sense insofar as its equations do not describe the behavior of its ultimate objects, such as molecules of a gas. It is, however, based on the realist assumption of an underlying nonstatistical multiplicity, whose individual members in principle conform to the causal laws of Newtonian mechanics. Ludwig Boltzman's distinguishability criterion, which gives meaning to the statement which individual constituent of a given multiplicity is which, is part of this classicality.

The denomination "realist" is usually further extended, as it is in this study, to theories or models that are approximate in this sense and, further, to theories that presuppose an independent architecture of reality that cannot be mapped, even partially or approximately, at a given point of history or possibly ever. Such architecture is, however, assumed to exist and possess attributes and structure by analogy with and on the *model* of the classical physical model. The latter is the main source of such conceptions, including when they are extended to quantum physics, for example, in Bohmian theories. In any event, in realist theories, a certain (structured) physical reality is assumed to be ultimately independent of observation, but is seen as subject to conceptual, theoretical, and mathematical, including quantitative, approximation, which physics can undertake in providing a mathematical theory of this reality. Such theories and the systems they describe or construct may be called *classical*, and one may speak, with Schrödinger (in his "cat paradox" paper), of the *classical ideal* in physics (Schrödinger 1935, *QTM*, pp. 152-153).

By contrast, quantum mechanics offers predictions, generally of statistical nature, concerning the system that *may not be* and, in certain interpretations, the present one among them, *cannot be considered as* causal or, more generally, subject to any realist conception. Indeed, as noted earlier, this suspension of realism, that is, the impossibility of assuming any conceivable underlying spatial-temporal dynamics governing quantum processes, obviously entails the lack of causality, since the latter is an assumed feature of such dynamics and hence is automatically disallowed. Schrödinger, we recall, stated this point by noting (with a very different evaluation of the type of argumentation offered here, which he saw as "born of distress") that, "if a classical state does not exist at any moment, it can hardly change causally" (Schrödinger 1935, *QTM*, p. 154). Even though the probabilistic predictions of quantum mechanics are subject to rigorous mathematical laws, in this case randomness and probability do not appear to arise in view of our inability to access the underlying causal dynamics determining, as in classical physics, the behavior of the system considered, in particular the behavior of individual quantum objects. It does not appear possible, and in the nonclassical interpretations such as Bohr's complementarity is rigorously impossible, to assume such a behavior to be causal or, again, to give it any conceivable form even at the level of idealization. This makes randomness and probability irreducible even and in particular in considering the behavior of individual quantum objects, such as electrons or photons (making it necessary to redefine such entities as well) or individual experimental events, the only entities (occurring at the classical macro-level) to which any quantum-mechanical predictions can apply.

2. RADICAL EPISTEMOLOGY AND IRREDUCIBLE PROBABILITY

Quantum phenomena (by which, again, I mean those physical phenomena in the analysis of which Planck's constant, h, cannot be treated as negligibly small) and quantum physics (by which I mean experiments and theories, such as quantum mechanics, dealing with quantum phenomena) appear to have made us confront the *fundamental, irreducible randomness* or *chance* and *irreducibly probability*.[47] Indeed, in dealing with quantum phenomena, we appear to have encountered or conceived of this type of randomness and probability for the first time in the history of science.

As we have seen throughout this study, a rigorous conceptual definition of what constitutes quantum phenomena themselves (including vis-à-vis quantum objects) is a

---

[47] I qualify this formulation because, as I explained, quantum mechanics or nature (either at the ultimate level of its constitution or in shaping our interaction with it) may ultimately prove not to be random and probabilistic in this radical sense, either by virtue of alternative interpretations of the available data or by virtue of new data. Karl Popper invokes in this context the notion of "objective probability," which, however, he never had sufficiently developed (Popper 1982). As noted in Chapter 2, Popper's views of quantum phenomena and quantum mechanics are very different from and even antithetical to those of Bohr.

nontrivial matter, and, in large measure, Bohr's life-long work on quantum mechanics was devoted to arriving at and refining such a definition. I shall return to the subject in the present context presently, reiterating for now that, following Bohr, phenomena are understood here in terms of manifest effects of the (quantum) interactions between quantum objects and measuring instruments upon certain parts of those instruments. These parts and the effects themselves in question are describable by means of classical physics, but are predictable (in statistical terms) only by means of quantum physics.

Thus, the probabilistic character of quantum-mechanical predictions appears to be due to the nature of our interactions with such (quantum) systems.[48] At the same time, however, the present interpretation does not assume and in fact prohibits positing an underlying causal architecture of reality that would be "distorted" by these interactions and would, thus, remain beyond our access in its actual form in full or even partial measure. It is the fact that such a causal architecture is assumed (usually by analogy with classical physics) that establishes the essential difference between the classical and the nonclassical view of probability, or reality. Throughout his writing, Bohr shuns and warns against the language of "distortion" as bound to lead to confusion in this respect. The interactions in question are defined, first, *technologically*, by the nature of the experimental technology that we use, which, as I said, defines all phenomena concerned, specifically all individual phenomena that could be subject to measurement and prediction. Secondly, these interactions are also defined, *phenomenologically* and *epistemologically*, by the fact that our capacity to conceive of the behavior of such systems is limited by the concepts of classical physics and by what might be called classical conceptuality in general, which we also continue to deploy, within certain limits, in quantum physics. It is this inapplicability of classical (and, thus, again any) concepts that makes the interpretation in question not only noncausal but, in the first case, radically nonrealist or nonontological (the lack of causality is, as I said, an automatic consequence of the impossibility of an ontological description at the quantum level).[49] Accordingly, in

[48] Quantum mechanics makes some exact predictions (i.e., within the capacity of our measuring instruments, which limitation also applies in classical physics), for example, as concerns the values of single conjugate variables involved in uncertainty relations, a fact used by Einstein, Podolsky, and Rosen in considering the EPR experiment. As explained in the preceding chapter, however, this statement requires further qualifications, which reveal the irreducibly statistical nature of quantum mechanics even in such cases.

[49] By "ontology" I do not mean the fact of the existence of "quantum objects" or of something to which this idealization might refer, which existence is obviously assumed by the present interpretation. Instead, by "ontology" I mean a possibility of providing an account, however indirect, or even forming a conception of the properties of the ultimate (here quantum) objects in question, including those properties that define

quantum mechanics the random character of the events considered and the apparently necessary, even in the case of individual phenomena, "recourse to probability laws" may be due, in Bohr's words, to "the inability of the classical frame of concepts to comprise the peculiar feature of indivisibility, or 'individuality,' characterizing the elementary processes" (*PWNB* 2, p. 34).

We are fortunate enough that, under these conditions of, to paraphrase and in part reverse Eugene Wigner's famous argument, the extraordinary *ineffectiveness* of mathematics in describing quantum objects themselves, rigorous statistical (and some determinate) predictions are possible, thus making quantum mechanics a rigorous physical theory in the same sense as classical physics is. Thus, mathematics is still as *effective* as ever in physics, and its effectiveness is, in this sense, all the more extraordinary, mysterious, in view of the fact that no underlying physical description, in mathematical or any other terms, of how the predicted data come about may be possible (it is, again, not possible on the present view). From this perspective, as discussed in Chapter 2, Heisenberg's matrix mechanics, which initiated the epistemology in question, also established a new kind of relationship between mathematics and physics. The terms of Bohr's statement on quantum probability cited above and its status as representing a particular view or interpretation of the situation in question require further explication, which I shall offer presently. First, however, and in part in order to ground this explication, I would like to cite another statement by Bohr on the nonclassical nature of probability in quantum mechanics. It opens one of Bohr's key works, the Warsaw Lecture of 1938, "The Causality Problem in Atomic Physics," where Bohr's concept of phenomena was introduced, as an extension of his analysis of the EPR argument of 1935 and related arguments by Einstein, considered in the preceding chapter. Bohr writes:

> The unrestricted applicability of the causal mode of description to physical phenomena has hardly been seriously questioned until Planck's discovery of the quantum of action, which disclosed a novel feature of atomicity in the laws of nature supplementing in such unsuspected manner the old doctrine of the limited divisibility of matter. Before this discovery statistical methods were of

---

them as objects in any conceivable sense, or of their behavior. One can, as I have done earlier, define "realist" more strongly so as to include "ontological" in this sense. Sometimes, however, it is also convenient to reserve the denomination "realist" for a more direct mapping or a more rigorous approximation (for example, on the model of classical physics) of the ultimate objects of investigation in question or their behavior. If one accepts the second view, one might, for example, see Bohmian mechanics as more ontological than realist, insofar as the actual *value*, or even actual *nature*, of variables defining the behavior of quantum objects is assumed but is never available. In any event, the present interpretation is neither ontological nor realist on either view. My argument would, accordingly, remain intact whether one distinguishes between "realist" and "ontological" in this way or not.

> course extensively used in atomic theory but merely as a practical means of dealing with the complicated mechanical problems met with in the attempt at tracing the ordinary properties of matter back to the behaviour of assemblies of immense numbers of atoms. It is true that the very formulation of the laws of thermodynamics involves an essential renunciation of the complete mechanical description of such assemblies and thereby exhibits a certain formal resemblance with typical problems of quantum theory. So far there was, however, no question of any limitation in the possibility of carrying out in principle such a complete description; on the contrary, the ordinary ideas of mechanics and thermodynamics were found to have a large field of application also to proper atomic phenomena, and above all to offer an entirely sufficient basis for the experiments leading to the isolation of the electron and the measurement of its charge and mass. Due to the essentially statistical character of the thermodynamical problems which led to the discovery of the quantum of action, it was also not to begin with realized, that the insufficiency of the laws of classical mechanics and electrodynamics in dealing with atomic problems, disclosed by this discovery, implies a shortcoming of the causality ideal itself. (*PWNB* 4, pp. 94-95)

Bohr, thus, makes a strong historical claim, which he elsewhere extends to philosophy as well. This extension is not surprising, given the fundamental relationships between general and physical classical conceptuality, since philosophical conceptuality may in turn be seen as a particular form of what we can conceive of in general. This view is clearly at work already in Kant or, in his confrontation with to Kant, Hegel and a long line of thinkers following both. The Kant-Hegel confrontation, indissociable from the question of classical physics, defines the history of modern philosophy, just as the Bohr-Einstein confrontation defines the history of quantum mechanics, from which philosophy is no longer dissociable in turn. Bohr says: "even in the great epoch of critical [i.e., post-Kantian] philosophy in the former century [centuries?], there was only question to what extent *a priori* arguments could be given for the adequacy of space-time coordination and causal connection of experience, but never question of rational generalizations [such as complementarity] or inherent limitations of such categories of human thinking" (*PWNB* 2, p. 65). Even the more radical philosophical questionings of causality, such those by Hume or Kant, are those of our epistemological capacity to perceive the underlying causal world, which would be presupposed at the ultimate level, in a kind of Platonist fashion, as inaccessible to us. It is only with Friedrich Nietzsche's and Charles Darwin's work (both, not coincidentally, contemporaneous with the emergence of thermodynamics) that the classical ideal of causality begins be questioned. In physics, one is compelled to agree with Bohr, it does not appear to happen prior to the discovery of quantum physics.[50] I would now like to sketch the quantum-mechanical concept of chance advanced here in general, philosophical terms. It is worthwhile to summarize the *classical* understanding of chance first.

Classically, chance or, more accurately (given that, classically, reality is always a causal necessity, and necessity reality), the *appearance* of chance is seen as arising from our insufficient (and perhaps, in practice, unavailable) knowledge of a total configuration of forces involved and, hence, of a lawful necessity that is always postulated behind an apparently lawless chance event. If this configuration becomes available, or if it could be made available in principle (it may not be available in practice), the chance character of the event would disappear. Chance would reveal itself to be a product of the play of forces that, however complex, is, at least in principle if not in practice, calculable by man, or at least by God, who, in this view, indeed does not play dice or at least always knows how they will fall. In other words, in practice, we only have partially available, incomplete information about chance events, which are, nonetheless, determined by, in principle, a complete architecture of necessity behind them. This architecture itself (it may also have temporal, dynamic aspects to it) may or may not be seen as ever accessible in full or even partial measure. The *presupposition* of its existence is, however, essential for and defines the classical view as causal and realist. On this point classical reality and classical causality come together; or rather this assumption of the ultimate underlying causal architecture of reality brings them together.

For example, if we cannot exactly (rather than only in terms of probabilities) predict how the dice will fall, or fully explain why a particular outcome has occurred, it is because the sum total of all the factors responsible is in practice unavailable to us. These factors may extend from a particular movement of a human (or perhaps divine) hand to minute irregularities in the material make-up of the dice themselves. In principle, however, a throw of dice obeys the laws of classical, Newtonian physics (or else chaos theory, which would not change the essence of the point in question). If we knew all such factors, we could predict and explain the outcome exactly by using these laws, which would describe both individual and collective behavior, and (law-fully) correlate them, in accordance with classical physical laws.

It is worth keeping in mind that, while a coin toss may be idealized as an individual event for the purposes of probability theory and may serve as a model for an individual event, for example, in quantum mechanics (say, for a photon going through or being reflected by a half-silvered mirror), it is really not. Any given coin toss, once viewed in terms of classical physics, is, at least in principle, subdividable (possibly all the way down to the level of the ultimate constituents of nature) into a sequence of events that are causal in character. The possibility, at least in principle, of this

---

[50] For a compelling account of this history, largely corroborating Bohr's view, see (Hacking 1984).

decomposition is what ensures the ultimate underlying causality of a coin toss. This is, in the present view, never possible for quantum-mechanical events or phenomena (in Bohr's sense), which can never be subdivided in this way. This cannot be done even in principle and even in idealized cases, which makes these phenomena both, and correlatively, truly individual and truly individually probabilistic, unlike classical events which are, thus, only collectively probabilistic.

Classical probability, whatever is the mathematics used to calculate it, masks causal sequences or networks of events. Subtle and complex as they may be, all scientific theories of chance and probability prior to quantum theory and many beyond it, and most philosophical theories of chance from the earliest to the latest are of the type just described, as Bohr observed in the statement cited earlier. They are causal (in presupposing the causality of the system they deal with) and most of them are also (and interactively) realist. Combined, two of Alexander Pope's famous utterances, the closing of Epistle 1 of *An Essay on Man* and his "Proposed Epitaph for Isaac Newton," encapsulate the classical view of chance and law:

> All Nature is but art, unknown to thee;
> All chance, direction, which thou canst not see;
> All discord, harmony not understood;
> All partial evil, universal good:
> And, spite of pride, in erring reason's spite,
> One truth is clear: Whatever IS, is RIGHT. (*An Essay on Man*, 289-94)

> Nature and Nature's laws lay hid in night;
> God said, let Newton be! and all was light. ("Proposed Epitaph for Isaac Newton, who died in 1727")

It is worth citing Einstein's version of this view of Newton, likening the light on the star of Newton to the actual stars:

> Seht die Sterne, die da lehren
> Wie man soll den Meister ehren
> Jeder folgt nach Newton's Plan
> Ewig schweigend seiner Bahn.

> [Observe the stars so you can learn
> How one ought to honor the Master:
> Each follows after Newton's plan
> Ever silently its path.][51]

---

[51]Cited by Banesch Hoffmann in (Hoffmann 1972, pp.141-42). The English translation (slightly modified here) is by Ivor Grattan-Guinness (Grattan-Guinness 1998, p. 260).

As we see the stars with the naked eye in the sky above us or insofar as classical physics applies, this is true and extraordinary. Once we deal with pulsars, subject to general relativity (Einstein's non-Newtonian theory of gravitation), Einstein himself becomes the Master, and with black holes one needs, along with general relativity, quantum theory and chance.

As here understood, the workings of nonclassical chance, such as that we encounter in quantum mechanics as complementarity, are fundamentally different from the classical picture just outlined. This chance is irreducible not only in practice but also and most fundamentally in principle. There is no knowledge in principle available to us, now or ever, that would allow us to eliminate chance and replace it with the picture of necessity behind it. Nor, however, can one postulate a causal dynamics (conceived, for example, by analogy or on the model of classical physics) as unknown or even unknowable but existing, in and by itself, outside our engagement with it. This qualification, correlative to the radical suspension of realism at the ultimate level of description, is crucial. For, as just explained, some forms of the classical understanding of chance allow for and are indeed defined by this type of (realist) assumption. The chance that we encounter in quantum physics is not only unexplainable in practice and in principle but is also irreducible in practice and in principle, irreducible to any (physical) necessity, knowable or unknowable, which view and the corresponding interpretation(s) make quantum mechanics involve, to return to David Bohm's language, certain irreducible lawlessness (Bohm 1995, p. 73). One could also speak, with Ole Ulfbeck and Aage Bohr, of "genuine fortuitousness" (Ulfbeck and Bohr 2001).

One might say that in classical physics there is only reality, however ultimately inaccessible, while chance is only apparent. By contrast in quantum physics (in the present interpretation) there are only probabilities of individual events without any underlying quantum-level reality or ontology. This reality or ontology is not only inaccessible but is unassumable. This, again, does not mean that nothing exists at that level or, again, the nothing exists in nature that is ultimately responsible for the situation thus theorized; quite the contrary, it is this existence that may ultimately be responsible for this situation, without allowing us to give this existence any conceivable form. The occurrence of any given individual effect is irreducibly lawless or fortuitous only in the sense, explained above, that the emergence of such an effect cannot be subject to a *physical* description or law by virtue of involving a quantum process. These effects themselves are subject to certain predictions of statistical and sometimes even determinate nature, for example, those concerning certain position or conversely momentum measurements (which, in the present interpretation would apply to measuring instruments). Either variable (but, in view of the uncertainty relations, never both together) could be predicted, at least by way of acceptable idealization, exactly, a fact that, as we have seen, was used by EPR in their argument.

I speak of "idealization" because this assertion requires a statistical qualification, since there is always a small but nonzero probability that our attempt to verify such predictions will not encounter the objects in question, as it would in classical physics. A single "measurement" cannot guarantee that a *relation* between a quantum object and the apparatus that has registered an event has in fact been established or even that an interaction with a proper (or any) quantum object has in fact taken place. In other words, a single event cannot guarantee that a proper measurement has in fact taken place (hence my quotation marks), anymore than an occurrence of a single spot of the screen in the double-slit experiment can guarantee that an emission from the source has in fact taken place. The occurrence of such "events" could only be ascertained on the basis of statistical considerations involving a sufficiently large number of experiments, whereby a certain particular type of correlations not found in classical physics is found. Accordingly, randomness is ultimately irreducible in quantum mechanics, at least in the present view, which, thus, offers us a new concept of chance, even if nature eventually proves not to conform to it.

Quantum mechanics is, in this interpretation, still a theory of individual events or effects, which is my main point here. Any analysis of such effects beyond their classical-like manifestation (including as concerns any possible subdivision) and beyond their irreducibly probabilistic nature is "*in principle* excluded," to return yet once more to Bohr's phrase that governed my argument in this study. Bohr's and the present view may be seen as conforming to what may be called *the individuality principle of quantum mechanics*. According to this principle, quantum mechanics is a theory of individual events rather than (only) collective events, or any other ensembles, even though and indeed because it must also be seen as an irreducibly probabilistic theory. This point was captured already by Born's rule for probabilistic predictions, which clearly refers to individual events, although he still appears to associate, albeit indirectly (as a "probability wave"), Schrödinger's ψ-function with a moving particle itself rather than seeing it strictly in terms of a predictive link between two classically manifest events or phenomena (in Bohr's sense). In the present view, all quantum-mechanical predictions concern only the outcomes (effects) of the interactions between quantum systems and our measuring instruments, the outcomes manifest in the classically observable parts of measuring instruments. These predictions give us no information whatsoever concerning the state of the quantum system itself under investigation either before, during, or after the measurement is performed, assuming that we can even speak of the physical state of such systems (e.g., in the spatial-temporal terms we use in classical physics). The possibility of excellent statistical predictions in the absence, at the level of individual objects, of not only all underlying causality but any assignable spatial-temporal dynamics is truly remarkable. How are these predictions possible? It may, I argue here, not be possible to ever answer this question.

Thus, while quantum mechanics, at least in the present interpretation, does not offer us laws that would in general enable us to predict with certainty the outcome of such individual events or when some of them might occur, it does require and depend on the concept of the individual physical event or phenomenon, which concept indeed defines quantum mechanics as quantum. By the same token the concept of a physical event or, correlatively, phenomenon is given a complex architecture in this interpretation, as explained earlier. As discussed earlier, this architecture was introduced by Bohr in the wake of the EPR argument and is defined by viewing such events as indissociable from the interaction between quantum objects and measuring instruments. These interactions become, it follows, irreducible in turn, in contrast to classical physics, where, as Bohr often stresses, they could, at least in principle, be neglected or compensated for. In other words, the individuality and indeed uniqueness of quantum-mechanical events appears and is defined at the level of phenomena, rather than that of quantum objects themselves or that of the dynamics of emergence of individual phenomenal effects, defined by the *quantum* interaction between quantum objects and the measuring instruments involved. It is worth reiterating that, on this view, while always unknowable, these dynamics are always different, unique and defined by (are irreducibly reciprocal with) the individual effects they lead to. This is also what allows us to speak of different fundamental entities, such as electrons or photons, even though, by the same token, we cannot always assign them identity or distinguishability at the quantum level. All quantum-statistical considerations and counting procedures, beginning with those involved in Planck's law, are based on the impossibility of so doing in certain circumstances, which poses apparently insurmountable difficulties in conceptualizing the situation but, self-evidently, poses no problem for the present view, since, according to it, no conception of any kind is applicable at this level.

Thus, as I have stressed throughout, while the (unavoidable) recourse to probability in quantum mechanics is an experimental fact, the *irreducibility of chance*, as understood by the present interpretation of the situation, is defined by the absence of an underlying architecture of reality that could in principle, if not in practice, be assumed to be causal. Indeed, according to this interpretation, no such architecture of any kind could, in principle, be assumed. In fact, under these circumstances, first, as Bohr argues, "there could be no question of attempting a causal analysis of [quantum] phenomena, but only, by a combined used of contrasting pictures, to estimate probabilities for the occurrence of the individual [quantum events]" (*PWNB* 2, p.34).[52] However, as Bohr adds, in accordance with the preceding argument: "It is most important to realize that the recourse to probability laws under such circumstances is essentially different in aim from the familiar application of statistical considerations as practical means of accounting for the properties of mechanical systems of great structural complexity. In fact, in quantum physics we are presented not with intricacies of this kind, but with the inability of the classical frame of concepts to comprise the peculiar feature of invisibility, or 'individuality,' characterizing the elementary processes" (*PWNB* 2, p. 34).

One should perhaps refer to the indivisibility *and* individuality (as we have seen, both features are found in it) of phenomena, arising by virtue of the elementary (i.e., quantum) processes. The classical frame of concepts can comprise neither these processes nor the emergence of these phenomena. Nor, accordingly, can it comprise the ultimate nature of the individuality and the indivisibility in question that. Nor, under these circumstances (i.e., given that no physical explanation of either quantum processes themselves or of the quantum interaction between quantum objects and measuring instruments is possible) could there be any question of a proper physical explanation or conceptualization for the application of probability laws, such as those reflected in Born's rule, von Neumann's projection postulate and the like. All such procedures are, in this interpretation, inevitably *ad hoc*. They are manifestations of the irreducible character of the quantum-mechanical chance and probability. As Bohr notes in his reply to EPR, yet again, bringing quantum epistemology and quantum probability together: "any comparison between quantum-mechanical and ordinary statistical mechanics,—however useful it may be for a formal presentation of the theory,—is essentially irrelevant. Indeed we have in each experimental arrangement suited for the study of proper quantum phenomena not merely to do with an ignorance of the value of certain physical quantities, but with the impossibility of defining these quantities in an unambiguous way" (*PWNB* 4, p. 78)

Thus, the present understanding of this character does not refer to any special general conception or mathematical theory of chance or probability, insofar as by these we refer to randomness of certain events or estimates of the occurrences of such events or their ranges. In this respect quantum chance is still chance (a random occurrence) and quantum probability is still probability (an estimate of how likely is an occurrence of a given event), just as they are in classical physics. This explains why Bohr did not appear to think that any special mathematical form of probability was necessary for quantum mechanics, anymore than any special form of quantum logic, of which he was even more suspicious (*PWNB* 3, pp. 5-6). The estimation procedures involved have their specificity, beginning with different ways of counting probabilities in quantum physics, defining

---

[52] I modify slightly this particular quotation, which, technically, comments on an earlier pre-quantum-mechanical state of quantum physics.

quantum theory from Planck on, which specificity, according to the present view, reflects the epistemology in question. In particular, as Bohr stressed, the rigorous impossibility of assigning individual identities to the entities of quantum multiplicities indicates how far we are here beyond any possible intuiting, conceptually or pictorially, their behavior, especially on the model of the classical, Democritean atomicity, which, as discussed earlier in this study, Bohr saw as inapplicable at that level or in quantum mechanics in general. This is partly why Bohr's interpretation no longer makes and ultimately prohibits this or any other assumption concerning the nature of quantum objects themselves and, as far as physics is concerned, speaks only of phenomena, in Bohr's sense, manifest in measuring instruments. From this point of view, one might say, as Bohr did in the passage on his concept of phenomenon cited above, that the situation is devoid of "any special intricacy" (*PWNB* 2, p. 64): we have well-defined individual physical phenomena and rigorous rules for predicting them. This resolution comes, it is true, at an epistemological price, exorbitant and even unacceptable to some, beginning with Einstein, of the impossibility of a further physical analysis of such phenomena and, hence, of the impossibility of physically justifying such rules.

It is also true that, in view of its specificity, the quantum-mechanical situation may require a delicate mathematical handling when we link to particular theories and conceptions of probability.[53] Part of the problem here may be the nature of probability theory itself as a mathematical theory and its own dependence on interpretation, arguably more significant than elsewhere in mathematics, specifically in conceiving certain aspects of the theory in terms of the occurrences of actual physical events. From this perspective, the Bayesian view of probability, especially in the context of quantum information theory, may hold a particular interest as far as the foundations of quantum mechanics are concerned. Significant as these questions may be, including as concerns the relationships between interpreting certain probability theories and interpreting quantum mechanics, I shall put them aside. They do not affect my main argument concerning the particular interpretation of quantum mechanics in question and the *physical* character of chance and probability it entails by virtue of its radically nonrealist epistemology. This situation appears to be unique in physics and perhaps in all science so far, but it also allows for a rigorous mathematical and theoretical treatment, even if only in terms of a particular interpretation, rather than in terms of an ultimate claim concerning nature.

As I explained at the outset, the view adopted in this study is defined by seeing Bohr's complementarity as only a particular interpretation of the situation that obtains in

---

[53] Such a handling may, for example, concern conditional probabilities and Kolmogorovism, as a number of theorists argue. I shall leave aside the question of physics involved in these arguments and certain further nuances concerning the differences between different interpretations (statistical, contextual, and so forth) involved in their arguments, since it would require an extended discussion of these articles, which cannot be undertaken here and is not essential to the present argument.

quantum mechanics and only in quantum mechanics (assuming the data currently available), rather than in other quantum theories, such as quantum field theories, let alone in nature itself. Accordingly, the irreducible inaccessibility of quantum objects and processes (including those leading to the effects of the interactions between quantum objects and measuring instruments) is assumed only to refer to the objects and processes as they are defined or, again, idealized by Bohr's interpretation. This remains the case, even though ultimately we may not be able to access them not only by means of quantum mechanics (in this interpretation) but also by any conceivable means. That is, this inaccessibility does not necessarily refer to what can ultimately be linked to such objects in nature, say, to the ultimate constituents of nature. Upon what happens at these levels as such, or how the situation can possibly be seen by other theories, this interpretation makes no claim, including the claim that it is inaccessible, unknowable, unrepresentable, inconceivable, untheorizable, undefinable, and so forth. It would be tempting to say in these circumstances (and one finds such claims) that, at the ultimate (quantum) level of the constitution of nature, there are no particles, no atoms, and so forth, rather than only that we cannot see the objects of quantum mechanics in these terms from the perspective of a particular interpretation. It seems more prudent, however, to adopt, as I have done throughout, the latter view, which is more cautious and which, in addition, suffices for the purposes of my argument here. As we have seen, it also follows that the ultimate or, as it were, the ultimate "ultimate objects" of quantum mechanics may be inaccessible even as inaccessible, unknowable even as unknowable, unrepresentable even as unrepresentable, inconceivable even as inconceivable, untheorizable even as untheorizable, and so forth. Such "entities" may not correspond in any way to the "objects" of quantum mechanics as here defined, even though and because this definition makes these objects unavailable to a description or conception in terms of the theory or in any terms, including "objects" or "entities." In other words, as I have stressed here, this view, too, must be seen as a (possibly ultimately inadequate) form of idealization.

This idealization is, however, not an approximation (of the kind classical theories often pursue), but is instead a kind of irreducible rupture from whatever may physically exist. Following and radicalizing Kant, we may not even be able rigorously to speak in terms of the existence in space and time or in terms of any specific form of materiality (particle-like, wave-like, or other) we can conceive of, since they may not be applicable, even in the sense of the remotest analogy. There may, as I said, be a link between such "un-objects" and the objects of the present interpretation of quantum mechanics, and what happens at these more remote levels may be ultimately responsible for everything at stake in quantum mechanics. If so, quantum mechanics involves two irreducible ruptures or discontinuities. The first is that between the knowable effects and the unknowable dynamics of their emergence, for which quantum objects and processes, conceived (as inconceivable) by this interpretation, are responsible. The second rupture is that between these objects, qua objects of the theory, and that—those un-objects—to which these objects can possibly be linked in nature. One of the consequences of this more complex, two-level, epistemological argumentation is that it leaves open space for other possible interpretations, possibly more classical in character. The validity and effectiveness of such alternative interpretations is of course a separate matter. I would argue that the present interpretation is, at least, as consistent and complete as any available. The present argument remains, following Bohr's but taking it to a still more radical epistemological limit, an argument for the completeness (an exhaustive account for the data) and consistency of quantum mechanics within its proper scope. Is this type of interpretation inevitable? If it eventually proves to be, one would not longer need to couple it to the interpretability of quantum mechanics, in the way I was compelled to do here, which would make my argument less cumbersome. It would be difficult to make a long-term guess, however. So, it may pay off to be cautious. For the moment, this interpretation is, at least, good enough, at least for quantum mechanics, and, as I shall argue, in the next chapter of this study, it may serve us well even beyond quantum mechanics and beyond Bohr.

Chapter 5. Complementarity, Quantum Mechanics, and Quantum Field Theory

1. BOHR , QUANTUM MECHANICS, AND QUANTUM FIELD THEORY: HISTORY AND PHILOSOPHY

This chapter pursues two related but distinct topics, one historical and the other theoretical, both of which are barely explored in the literature on Bohr and on quantum theory. The first is the significance of quantum field theory for the history of quantum mechanics, including Bohr's work, and the debates concerning it. The second is the relationships, especially philosophical and epistemological, between quantum mechanics and quantum field theory.[54] A sustained exploration of these topics and the relationships between them is beyond the limits of this study. I thought it desirable, however, and in some respects imperative, to offer a sketch of the subject, first, in order to give a proper closure to my argument and, second, in order to explain why, as indicated at the outset of this study, quantum field theory leads us beyond Bohr but without leaving him behind. Bohr himself saw quantum electrodynamics and quantum field theory as confirming his key ideas concerning the epistemology of quantum mechanics, and possibly extending them through the new mathematics and physics that they help to develop.[55]

---

[54] It may be useful to note for the non-specialist reader that in quantum field theory the effects of Einstein's (special) relativity theory are taken into account, which make the theory relativistic, in contrast to quantum mechanics, where such effects can be neglected because the speed of the objects considered by it is slow vis-à-vis the speed of light. As discussed in Chapter 3, as a local theory, quantum mechanics remains *consistent* with relativity as concerns all predictions it makes. In question here is the difference in scope between both theories.

[55] The philosophy of quantum field theory is a complex and, even though the theory itself, at least quantum electrodynamics, has been around in pretty much its current form for half a century, still little developed subject, in comparison with quantum mechanics, as the paucity of literature addressing it, as against that on quantum mechanics, suggests. One can think of barely a handful of books devoted to the subject vs. dozens, if not hundreds, of books on the philosophy of quantum mechanics. One might mention in particular Paul Teller's *Quantum Field Theory: An Interpretive Introduction* (Teller 1995), which also contains useful further references. For helpful historical accounts see Silvan Schweber's *Quantum Electrodynamics and the Men Who Made It* (Schweber 1994) and Steven Weinberg's (more technical) *The Quantum Theory of Fields, Volume 1: Foundations* (Weinberg 2005). An essay offered in this chapter

My historical argument is more immediately and more easily made, even without offering a detailed discussion of the subject, which is, again, beyond my scope but which, I would contend, would strengthen the case I want to make. The birth of quantum electrodynamics and quantum field theory was nearly simultaneous with that of quantum mechanics. In some respects, the former even preceded the latter. It is nearly certain (and assumed by most commentators) that Schrödinger wrote a relativistic wave equation for the electron before discovering his nonrelativistic equation. The former equation is now known as the Klein-Gordon equation and is used elsewhere in quantum field theory, while the relativistic (free) electron is described by Dirac's equation. Classical electrodynamics was of course crucial to Planck's work on his black body radiation law, which gave birth to quantum physics. The discussion of quantum field theory appeared already in the so-called three-man paper on matrix mechanics by Born, Heisenberg, and Jordan (1926). Most especially, however, I have in mind the early work on quantum electrodynamics by Jordan (primarily responsible for the discussion of the subject in the three-man paper) and Dirac (who is widely seen as primarily responsibly for founding the theory), and then Dirac's discovery of his relativistic equation for the electron. Both Jordan and Dirac were also co-founders of the so-called transformation theory, which connects Heisenberg's and Schrödinger's versions of quantum mechanics. Dirac's discovery of his equation was itself an example of his masterful use of the transformation theory, his "darling," as he called it. It is, accordingly, not surprising that quantum electrodynamics and quantum field theory had a much greater impact on the debates concerning quantum mechanics from the outset and then throughout its history than previously explored or even acknowledged. This includes the work on the subject by both Bohr and Heisenberg.

By the early 1930s, Heisenberg saw quantum mechanics as what he called a *closed* theory or, in Thomas Kuhn's idiom, paradigm, similarly to the way classical physics was closed within its proper limits (Kuhn 1962). This need not mean, and was not by either Heisenberg or Kuhn, that either theory was simply completed, but instead that they were both paradigmatically defined in their fundamental aspects, or, in Kuhn's terms, were operating in the regime of "normal" rather than (as during the period of their emergence leading to a paradigm shift) "revolutionary" science. Heisenberg thought, in many respects, rightly, that a new paradigm was introduced with Dirac's discovery of antimatter (a consequence of Dirac's relativistic equation for the electron), which he considered as "perhaps the greatest of all great changes" in twentieth-century physics. This assessment is not often cited and is even more rarely appreciated (not in recent years in any event), in part because his reasons for making it are rarely adhered to. These

---

could only make a limited contribution, historical or philosophical, to the subject, but, hopefully, a helpful one, to both the specialist and the non-specialist reader, by virtue of relating it to Bohr's work and the epistemological problematic discussed in this study.

reasons, which I shall discuss in detail below, reflect profound physical and epistemological questions raised by quantum electrodynamics and quantum field theory, including vis-à-vis quantum mechanics.

After quantum mechanics was established, at least in physical terms, if not as its interpretation was concerned (this process had taken much longer and is continuing even now), Heisenberg's main interests appear to have shifted to quantum field theory. He played a pivotal role in establishing it, in particular, it is worth noting, as a theory of fields rather than particles, as it was in Dirac's work, and also as a gauge theory (now, the dominant form of quantum field theory), in his collaborations with Pauli in the late 1920s. I shall comment on the concepts of particles and fields in quantum field theory in the next section, and I shall only mention here that, historically, quantum electrodynamics, the original form of quantum field theory, was introduced by Dirac in 1926-1927 as a theory in which both electrons and photons were treated in terms of particles. By contrast, Heisenberg and Pauli developed, by analogy with or by quantizing classical electrodynamics (a classical field theory of electromagnetic radiation, introduced by James C. Maxwell), in which both electrons and photons were treated in terms of fields (not the same as waves, although both concepts are linked in classical electrodynamics). Or more accurately, given that the quantum nature of electron and photons was retained, they were treated in mathematical terms analogously to the way fields are treated in classical electrodynamics. A little later, in 1930, Enrico Fermi introduced a version of quantum field theory in which the photons were treated in terms of fields and the electrons in terms of particles. Heisenberg continued to make major contributions to it for the decades to come, in some respects almost as significant as his contributions to quantum mechanics, his work on the *S*-matrix formalism arguably the most significant among them. So did Pauli, who in 1953 virtually co-discovered (but never published) the Young-Mills theory, which eventually became the grounding theory for the so-called standard model, which defines the current state of elementary-particle physics.

By contrast, although he made some important contribution to nuclear physics, Bohr, for the rest of his life, continued to work on and refine, in part under the impact of Einstein's arguments, the philosophical and epistemological architecture of complementarity as an interpretation of quantum mechanics. As a result, he continued to shape decisively and even uniquely the debate concerning quantum mechanics and its interpretation. While Heisenberg's thinking and writings (technical, philosophical, and popular) made their impact on this debate, his contribution to the interpretation of quantum mechanics was relatively limited and in large measure followed that of Bohr, as he himself expressly stated on several occasions. Certain differences between his views and those of Bohr, specifically as concerns the role of mathematics in quantum theory, are noteworthy and had their impact throughout the history of quantum theory. On the other hand, as I discussed earlier in this study, Bohr saw the role of mathematics of quantum mechanics, or of quantum field theory, as more significant and more in accord with Heisenberg's view of this role than is customarily thought or might appear.

Quantum electrodynamics and quantum field theory, most specifically Dirac's work (but also that of Pauli and Heisenberg), retained their significance for Bohr's thinking and his work on complementarity throughout his life. Both were given major attention in Bohr's Institute in Copenhagen. Dirac's first paper on quantum electrodynamics was written, while in Bohr's institute in Copenhagen in 1926, that is, at the time of Bohr's work leading to complementarity and Heisenberg's work leading to the uncertainty relations, and their famously heated exchanges on quantum mechanics and its interpretation. Several among major physicists working there, who also assisted Bohr in his work, such as Heisenberg or, earlier, Hendrik Kramers, and then Oscar Klein and Léon Rosenfeld, made major contributions to quantum field theory. Most of the key works by Bohr, beginning with the Como lecture, contain important references to quantum electrodynamics and quantum field theory. An important exchange of letters between Bohr and Dirac occurred at the time of Dirac's work on his positron theory. Dirac's letter announces ideas that have had a lasting impact on the subsequent development of quantum theory. Most significant perhaps is Bohr and Rosenfeld's collaboration on the question of measurement on quantum field theory in "On the Question of Measurability of Electromagnetic Field Quantities" (1933) and "Field and Charge Measurement in Quantum Electrodynamics" (1950), the first of which was written in response to Lev Landau and Rudolf Peierls's argument (*QTM*, pp. 479-522 and pp. 523-34). The exchange, which has preceded EPR's argument and Bohr's reply only by two years, involved crucial physical and epistemological issues concerning the possible differences between quantum mechanics and quantum electrodynamics or quantum field theory, in particular the question of the uncertainty relations, which Landau and Peierls claimed to be no longer applicable in quantum field theory. Bohr and Rosenfeld argued to and demonstrated the contrary. Bohr was actively involved and continued to exert a significant influence in the discussions concerning earlier stages of quantum field theory, including as then emerging theory of nuclear forces, in the early 1930s.[56] Accordingly, it would be difficult to think that quantum field theory did not affect his work on the epistemology of quantum mechanics and complementarity, during this period, in particular his response to EPR, and in the subsequent years. The development of quantum field theory has offered Bohr continuing support for his

---

[56] A. Pais's discussion of quantum field theory and nuclear physics during the 1930s gives a clear sense of Bohr's engagements with it and his influence on its development (Pais 1986, pp. 296-438), as does his discussion of this part of Bohr's work in the corresponding parts of his biography of Bohr (Pais 1991, pp. 346-374). The title of this chapter is, interestingly, "Toward the edge of physics in Bohr's style, and a bit beyond." One may, as I argue here, need to move more than "a bit beyond" Bohr in approaching the edge of physics, even without leaving him behind.

argument concerning the epistemology of quantum phenomena and quantum theory, quantum field theory included.

As a quantum theory, quantum field theory appears to retain most of the key physical, mathematical, and epistemological features of quantum mechanics as considered in this study, which features define both as *quantum* theories. If one could speak of a single conceptual feature that defines the difference between them, it would be that of the creation and annihilation, birth and disappearance, of particles from the false vacuum, and, correspondingly, of the virtual particle formation. This concept may also be seen correlative to the role the concept of (quantum) field itself plays in quantum field theory, as against quantum mechanics, and gives it its name. One could see at least some of the *concepts* of both quantum mechanics and quantum field theory (of course of classical physics) as *philosophical* concepts. According to Frank Wilczek, a leading contemporary quantum-field theorist and a Nobel Prize laureate, "the primary goal of fundamental physics is to discover profound concepts that illuminate our understanding of nature" (Wilczek 2005, p. 239). The concepts Wilczek has primarily in mind are physical and, it is worth noting, quantum-field-theoretical concepts, such as gauge invariance, symmetry breaking, the Higgs field, or superconductivity, and they must be, given the disciplinary character of modern physics as a mathematical-experimental science. These concepts may, however, also be seen as, or as involving, philosophical concepts, as Bohr's concepts, such as complementarity, phenomena, or atomicity (all carried over into quantum field theory, at least on Bohr's or the present view) especially, but not uniquely, exemplify.

As I noted above, Bohr saw quantum electrodynamics and quantum field theory as confirming his key ideas concerning the epistemology of quantum physics, as he elaborated them in the case of quantum mechanics, and possibly as requiring a further radicalization of those ideas, primarily in the same direction, that is, away from classical physics. He described Dirac's theory as "a most striking illustration of the power and fertility of the general quantum-mechanical way of description" and as reflecting "new fundamental features of atomicity" (*PWNB* 2, p. 63). The term "atomicity" is clearly used here in Bohr's special sense, as discussed in this study, since the comment occurs in "Discussion with Einstein," largely defined by this concept. In other words, quantum field theory moves beyond quantum mechanics ("new fundamental features") without leaving it behind, as it retains "the power and fertility of the *general* quantum-mechanical way of description" and its physical and philosophical concepts, such as "atomicity," and, thus, one might add, it moves beyond Bohr, without leaving him behind.

This question—that is, to what degree higher level quantum theories such as quantum field theory follow the epistemology of quantum mechanics or depart from it either by extending it more radically or, conversely, by bringing physics closer to classical physics—is my second main concern in this chapter. My argument follows Bohr's view and Heisenberg's insights in the wake of and responding to Dirac's discovery of antimatter: quantum electrodynamics and quantum field theory contain new epistemological complexities that may and are likely to take them further in the direction initiated by quantum mechanics. Indeed these complexities possibly make them, *as currently constituted*, depart as much or even more from quantum mechanics than the latter does from classical physics. I qualify my assessment because it is possible that these theories will eventually be developed in an epistemologically more classical direction or replaced by other theories that are epistemologically more classical even within their present scope, that is, even apart from the fact that these theories are manifestly incomplete by virtue of not incorporating gravity. The standard model, accounting for electroweak and nuclear forces, is not yet quite finalized or unified either. This chapter, thus, and with it this study, ultimately move beyond Bohr, although they also argue that Bohr or at least reading Bohr may still be the best way to do so. My argument, however, also makes apparent, with the help of quantum field theory, that moving beyond Bohr need not leave Bohr behind, and may indeed also be the best way to read and to understand Bohr.

2. CREATION AND ANNIHILATION OF PARTICLES: "PERHAPS THE BIGGEST OF ALL THE BIG CHANGES IN PHYSICS IN OUR CENTURY"

I would like to clarify first the use of the concepts of particle and field in quantum field theory. As indicated earlier, the theory could be presented mathematically and experimentally (predictively) equivalently either in terms particles or in terms of fields, a concept developed by Faraday and Maxwell in the nineteenth century in order to handle electromagnetic radiative phenomena, such as light.

We recall that, as discussed in Chapter 1, in developing quantum mechanics, first Heisenberg and then Born and Jordan, formally adopted the equations that would describe the motion of particles in classical mechanics, most generally, in their Hamiltonian form, but gave them a new, quantum-mechanical, physical content by using different (matrix) type of variables to which the equations themselves applied. This was also a very different *type* of physical content, along with and, correlatively, to a very different *type* of mathematics that results, as against those of classical physics. In particular, on the present view, physically, quantum mechanics only predicts, and moreover, in general, only in terms of probabilities, the outcomes of the experiments it consider, but does not describe the properties of quantum objects and their behaviors in the way classical mechanics describes the behavior of classical physical objects. (The determinism of classical predictions arises by virtue of the possibility of such a description.) Mathematically, the theory used a new type of variables, infinite matrices or, equivalently, operators in Hilbert spaces (over complex numbers), as differential functions of real variables used in classical physics. This was the meaning or, at least, the import of Heisenberg "new kinematics."

In developing his quantum electrodynamics, in which, as I noted, both electrons and photons were seen in particle terms, Dirac adopted the same type of approach, which he used just a few months earlier in creating his own version of quantum mechanics, using Heisenberg's original paper, but independently of Born and Jordan. In the case of quantum electrodynamics, he developed a more complicated Hamiltonian formalism applied to the type of matrix or operator variables that are analogous (but not quite identical) to that used in quantum mechanics. By contrast, in their version of quantum field theory Heisenberg and Pauli used the field picture. In this respect is was the first rigorous quantum *field theory*, since, while a form of electrodynamics, as a particle theory, Dirac's theory could be seen as still a form of *mechanics*. Heisenberg and Pauli's approach was more similar to that of Schrödinger in developing his wave mechanics, but, unlike Schrödinger, Heisenberg and Pauli, suspended the physical picture of wave propagation, or all classical-like physical pictures, from the outset. They used the equations analogous to the (wave) equations of classical electromagnetism, Maxwell's famous equations, to develop their mathematical theory. (Schrödinger's equation itself could be seen in "particle-like" terms, the point noticed, with some surprise, by Schrödinger himself, who was also one of the first to discover the mathematical and predictive equivalence of his wave mechanics and matrix mechanics, the equivalence manifest in this possibility of seeing his equation in either terms.) The variables used in Heisenberg and Pauli's equations were, again, different from those used in classical electrodynamics, and, as those of quantum mechanics, these equations only predict the probabilities of the outcomes of the experiments in question. Later on Heisenberg, Fermi, and Hideki Yukawa introduced quantum field theory of nuclear forces.

These events inaugurated quantum field theory as a theory of particles and fields, which it has remained ever since, allowing one to use either picture or variously combine both, depending on one's need or preference. Both types of approach work, but with different effectiveness in different circumstances. In this respect it is analogous (but not identical) to the way one could use either Schrödinger's "wave" equation or a more algebraic (with qualifications given in Chapter 2) formalism of matrix mechanics or operators in Hilbert spaces. The complementarity of particles and fields is sometimes invoked as well. It is, however, different from the wave-particle complementarity of quantum mechanics, and requires an even greater caution in using any physical analogies with classical physics. One might say that the complementarity of particles and fields is more mathematically defined, although, on the present view, neither complementarity, nor the concepts of particles, waves, or fields, to begin with, are rigorously applicable to quantum objects themselves or their behavior. Bohr never invokes the complementarity of particles and fields. But then, as discussed earlier, he does not really use the wave-particle complementarity either. The primary reason in both cases appears to be that neither complementarity really corresponds to the mutually exclusive individual situations of measurements or phenomena in Bohr's sense.

In both quantum mechanics and quantum field theory alike, however, we deal with two or more equivalent mathematical models, mathematically analogous to those *describing* the behavior of two irreducibly different entities, *either* particles (mechanics) *or* waves (field), in classical physics. Neither such models nor, obviously (i.e., given that an actual object cannot be simultaneously continuous and discrete, point-like), such entities themselves can be combined in classical physics. By contrast, in either quantum mechanics or quantum field theory, the two models in question (these models, again, are not the same as in classical physics) are now used only for *predicting* certain aspects of the behavior of the *same* entities, quantum objects, which are neither particles nor waves, or possible nothing else we can conceive of. All predictions and hence all probabilities involved in quantum field theory concern only the effects of the interactions between quantum objects and measuring instruments upon the latter, some of which effects could be seen collectively wave-like, although composed of discrete individual entities or phenomena. This view, as we have seen, allows Bohr to avoid paradoxes involved in jointly attributing such incompatible properties of waves, on the one hand, and particles, on the other, to quantum objects, which are never observed together, in the first place, the fact of which, as we have seen, both Bohr and Heisenberg took advantage.

Quantum field theory, thus, retains the epistemological features of quantum mechanics as complementarity, or at least it may and perhaps must be interpreted accordingly, at least for now. Indeed, both quantum mechanics and quantum field theory describe the same quantum objects in the first place, but in different circumstances, defined by the levels of energy at which the corresponding processes take place. Mathematically, too, the Hilbert-space formalism (over complex numbers), arguably the dominant form of quantum-theoretical-formalism (even when other theories, e.g., group theory, are deployed) is used, albeit differently in some important respects, in both cases, and may be seen as defining all of quantum theory vis-à-vis classical physics. Quantum field theory is, however, conceptually different from quantum mechanics insofar as it adds, the concept of virtual formation, creation and annihilation, of particles, which introduces new complexities and stratifications into the conceptual architecture of the theory, vis-à-vis quantum mechanics, even as it retains the key, including the most radical, epistemological aspects of the latter. Here, and for the remainder of this chapter, I adopt the particle picture, while keeping in mind the qualifications just offered. This new concept of virtual particle formation is also reflected in the difference in the mathematical structures of quantum mechanics and quantum field theory, although in both cases we, again, deal with Hilbert spaces. These difference initially emerged in the so-called second quantization, which was introduced by Jordan in the three-man paper and which replaced a one-body or one-degree-of-freedom problem with a many-body problem, technically, the infinitely-many-body or infinite-degree-of-freedom problem. On this view, in Abraham Pais's words, "the hydrogen atom can no longer be considered to consist of just one proton and one electron. Rather it contains infinitely many particles. … A transition from one [energy] level to another must therefore mean that particles with energy $h\nu$ are either made or else disappear' (Pais 1986, pp. 325, 333). This conception was eventually extended and properly developed in terms of the mathematical formalism (and experimentally confirmed) for to all fundamental forces (except, again, gravitation) and led to what became know as the standard model.

I shall now explain, in elementary or "naïve" (but, I would contend, essentially correct) terms, why one is compelled to move in this direction, and, I also argue, expand or add further stratification and structure to the radical epistemology of quantum mechanics discussed in his study. It does so especially by introducing a more complex and more radical form of multiplicity in our picture of physical phenomena, including in Bohr's sense of the term, epistemologically inevitable here. It may not be possible to radicalize this epistemology further insofar as it makes all the ultimate object and processes in question in all quantum theory irreducibly unrepresentable or even inconceivable by any means available to us, including as "objects" and "processes."

Suppose that one arranges for an emission of an electron from a source and then performs a measurement at a certain distance from that source. Merely placing a photographic plate at this point would do. The corresponding traces could, then, be properly treated by means of quantum field theory, which, as does quantum mechanics, deals, in terms of statistical predictions, with such traces of events, or such events defined by traces, and only with them, rather than offers a mathematical description of the behavior of quantum objects in the manner of classical physics.

First, however, let us assume, along the lines of the old, pre-quantum-mechanical, quantum theory, that we deal with as a classical physical object. According to classical physics, one would encounter at this point the same object, and its position could be predicted exactly by classical mechanics.

In quantum mechanics, by contrast, one would encounter either an electron or nothing, and quantum mechanics predicts the alternative probabilities for such events, for example, fifty percent for each. This is why, as explained in Chapter 4, chance and probably are unavoidable in quantum mechanics.

Once the situation involves higher energies and is governed by quantum electrodynamics, the original form of quantum field theory, one may find an electron, nothing, a positron (anti-electron), a photon, an electron-positron pair, or, once we move to still higher energies or different domains governed by quantum field theory, still something else. What is referred to by the absence of a single ground state in quantum field theory is a manifestation of the situation. Just as does quantum mechanics, quantum field theory rigorously predicts which among such events can or cannot occur, and with what probability. And as in quantum mechanics, the probabilities, and only probabilities, for the alternatives are properly predicted by quantum field theory, which makes chance and probability as unavoidable in quantum field theory as they are in quantum mechanics.

The upshot is that in quantum field theory, an investigation of a particular type of quantum object (say, electrons) not only irreducibly involves other particles of the same type but also other types of particles, conceivably all existing types of particles. It is important that the situation involves *different types* of particles, since the identity of particles within each type is strictly maintained in quantum field theory, as it is in quantum mechanics in fact. Indeed, it is crucial for quantum field theory, or quantum mechanics, that one cannot distinguish different particles of the same type, such as electrons, and, accordingly, one can never be certain that one encounters the same electron in the experiment just described even in the quantum-mechanical situation.

Thus, it is as if instead of identifiable moving objects and motion of the type studied in classical physics, we encounter a continuous emergence and disappearance, creation and annihilation, of particles from point to point, theoretically governed by concept of the *virtual* particle formation. This view takes us even beyond quantum mechanics, which has already radically put into question such classical physical concepts as objects, either particles or waves, and motion. Nobody has ever *seen* a moving quantum object, such as an electron or a photon, even with the help of measuring instruments, however powerful, which is, we recall, one of the starting point of Heisenberg's founding work on quantum mechanics. Or rather, while retaining the radical epistemology of quantum mechanics, in particular the lack of causality and realism due to the ultimately inconceivable nature of the ultimate objects and processes in question, quantum field theory complements them by the concept of creation and annihilation of particles and, correlatively, the concept of virtual particle formation.

The corresponding operators, introduced by the so-called secondary quantization of Jordan and Dirac (there certain differences between both approaches), that is, the operators used (analogously to those of the standard quantum mechanics) to predict the probability of such events, are called the creation and annihilation operators. The introduction of these operators and, with them, a new mathematical formalism (although still of a Hilbert space type, but involving more complicated Hilbert spaces) was a momentous event in the history of quantum physics, comparable to that of Heisenberg's introduction of his matrix variable, essentially also Hilbert-space operator variables, in the case of quantum mechanics. Both of these inventions may be compared to Newton's introduction of his function variables and, with them, differential calculus in the case of classical physics. In all three cases, we encounter an invention of a new calculus, which also links mathematics and physics, and defines, and in each of these cases, redefined the relationships between them.

To return to Bohr's assessment of Dirac's theory, cited earlier, prompted by these considerations: "Dirac's ingenious quantum theory of the electron offered a most striking illustration of the power and fertility of the general quantum-mechanical way of description. In the phenomena of creation and annihilation of electron pairs we have in fact to do with new fundamental features of atomicity, which are intimately connected with the non-classical aspects of quantum statistics expressed in the exclusion principle, and which have demanded a still more far-reaching renunciation of explanation in terms of a pictorial representation" (*PWNB* 2, p. 63). It is, again, not quite clear how much further one can renounce "explanation in terms of a pictorial representation," once such a representation or indeed any representation or even conception are already renounced altogether, as they are in quantum mechanics as complementarity. It is true, however, that Dirac's theory and, following it, all quantum field theory as now constituted, makes a return to classical-like epistemology all the more unlikely (assuming, as some do, that such a return is still possible in quantum mechanics), but instead introduces new levels of stratification and structure into quantum theory. "Atomicity" should be understood here in Bohr's sense, as defined by him earlier in the same article, "Discussion with Einstein," and, as discussed earlier in this study, governing his view of "the general quantum-mechanical was of description," including in quantum field theory. In other words, we are still dealing here with the effects of the interactions between quantum objects and measuring instruments upon those instruments, but in the case of quantum field theory, with a greater multiplicity of such effects and more complex rules for probabilities for the events (effects) involved. It is clear, in particular, that each actual, registered event in the quantum-field-theoretical picture given above could and, it appears, must be presented in terms of effects, and hence, indivisible phenomena or "atoms" in Bohr's sense, while "virtual" events relate to possible phenomena.

Heisenberg held similar views. He was undoubtedly aware of Bohr's view of the subject and knew well Bohr's statements here cited, but, in this case, he did not need Bohr to assess the situation. Indeed, as noted earlier, his saw Dirac's theory as an even more radical revolution than quantum mechanics was, at least in physical and mathematical terms, but also, now consistently with Bohr, in epistemological terms, insofar, as the theory introduced more radical forms of multiplicity, if not that of unknowability or inconceivability, into our picture of physical phenomena, in either sense. For, as just explained, an appeal to Bohr's sense of phenomena appears as unavoidable here as in quantum mechanics. In reflecting on the quantum-field-theoretical situation just explained in an essay revealingly entitled, "Development of *Concepts* in the History of Quantum Mechanics" (emphasis added), in early 1970s, Heisenberg was, as I noted above, compelled to see Dirac's discovery of anti-particles, correlative to this situation, as "perhaps the biggest of all the big changes in physics of our century" (emphasis added). He explained: "It was a discovery of utmost importance because it change our whole picture of matter" (Heisenberg 1983, pp. 31-32). He elaborated as follows:

> So the final result at this point seems to be that Dirac's theory of the electron has changed the whole picture of atomic physics. After abandoning the old concept of the elementary particles, those objects which had been called

"elementary particles" have now to be considered as complicated compound systems, and will have to be calculated some day from the underlying natural law, in the same way as the stationary states of complicated molecules will have to be calculated from quantum or wave mechanics. We have learned that energy becomes matter when it takes the form of elementary particles. The states called elementary particles are just as complicated as the states of atoms and molecules. Or, to formulate it paradoxically: every particle consists of all other particles. Therefore we cannot hope that elementary particle physics will ever be simpler than quantum chemistry. This is an important point, because even now many physicists hope that some day we might discover a very simple route to elementary particle physics, as the hydrogen spectrum was in the old days. This, I think, is not possible. (Heisenberg 1983, pp. 34-35)

The theory made a remarkable progress and has acquired a much richer content and structure since its introduction or even since Heisenberg's remark, as is manifest most famously in the electroweak unification and the quark model of nuclear forces (although these development commences around the time of these remarks and Heisenberg was certainly aware of them). The essential epistemological point in question, however, has remained in place, just as is the case with the epistemology of quantum mechanics, and in this respect Bohr's and Heisenberg's statements require only relatively minor adjustments. A more recent statement by Pais, himself a major practitioner of the theory and a major historian of the subject, confirms that epistemologically we still confront the same situation. The statement is, it is true, no longer that recent either. It was made in 1986, but not much changed in this respect since, as is clear from Wilczek's article, cited above, which dates to 2005 and thus is truly recent (Wilczek 2005). In epistemological and philosophical terms, the picture of the current state of particle physics in terms of quantum field theory given by Wilczek's article conforms to the one given by Heisenberg, and many of the key concepts, such as elementary particles, symmetry (and elementary particles as symmetry), and so forth are much the same. The actual achievements of the theory (or relevant experimental physics) during the last decades, those of Wilczek's own work among them, were, again, momentous, in part by given new specific mathematical and physical content to these concepts, but also by giving a new physical content to the mathematical formalism. Indeed it was also quantum field theory that led to string and then "brane" theories, the current stratosphere of theoretical physics.[57] Certain new concepts were added as well,

---

[57] A remarkable and remarkably efficient (in two pages!) discussion of the essential and immediate nature of these connections is offered by Anthony Zee (Zee 2003, pp. 38-40). This discussion is highly technical, however, although reasonably elementary by the current standards of theoretical physics, especially those of string or

building on the conceptual framework under discussion here, including again in Wilczek's work, such as the so-called "confinement" or "asymptotic freedom," defining the behavior of quarks, or string and branes, all of which have a complex physical, mathematical, and philosophical content. According to Pais:

> Is there a theoretical framework for describing how particles are made and how they vanish? There is: quantum field theory. It is a language, a technique, for calculating the probabilities of creation, annihilation, scattering of all sorts of particles: photons, electrons, positrons, protons, mesons, others, by methods which to date invariably have the characters of successful approximations. No rigorous expression for the probability of any of the above-mentioned processes has ever been obtained. The same is true for the corrections, demanded by quantum field theory, for the positions of energy levels of bound-state systems [e.g., atoms]. There is still a [Schrödinger] equation for the hydrogen atom, but it is no longer exactly soluble in quantum field theory. In fact, in a sense to be described [i.e., the sense explained above], the hydrogen atom can no longer be considered to consist of just one proton [or three quarks plus gluons in the nucleus] and one electron. Rather it contains infinitely many particles.
>
> In quantum field theory the postulates of special relativity and of quantum mechanics are taken over unaltered, and brought to a synthesis which perhaps is not yet perfect but which indubitably constitutes a definitive step forward. It is also a theory which so far has not yielded to attempts at unifying the axioms of general relativity with those of quantum mechanics. Is quantum field theory the ultimate framework for understanding the structure of matter and the description of elementary processes? Perhaps, perhaps not. (Pais 1986, p. 325)

All of these features of quantum field theory are still in place. In the present view, one would not be able to speak rigorously of "the *description* of elementary processes," as this inability would itself appear as a postulate of quantum mechanics in the present interpretation taken over unaltered by quantum field theory. Pais does not make strong claims to that effect in the book either. It is worth noting that Pais was Bohr's assistant in 1940s and wrote a more or less definitive biography of Bohr (Pais 1991), as well as that of Einstein (Pais 1982), and he appears to be close to Bohr's philosophy of quantum physics.

It may be added that the complexities of quantum field theory discussed here and summed up by this statement inhabit the physical facts that are stated by such

---

brane theories. For an *elegant*, if somewhat over-enthusiastic, popular account, see Brian Greene's *The Elegant Universe: Superstrings, Hidden Dimensions, and the Quest for the Ultimate Theory* (Greene 2003).

simplified assertions that the hydrogen atom consists of one proton and one electron, that the proton is composed of three quarks (with some gluons linking them), and so forth. These complexities are sometimes bypassed or even misrepresented in popular expositions of the subject, including by physicists, and such expositions can, accordingly, be misleading in this respects, even at a naïve level. One rarely encounters a kind of clear and clarifying picture given by Pais here and elsewhere in his book, although much of the book would not be accessible to a lay reader. Some of it is, however, and is well worth the effort. The mathematical complexities involved are often prohibitive, even for many physicists (who are not in the field), and require literally decades of training.

In the present-day practice of quantum field theory quantum-field-theoretical events are represented in terms of Feynman diagrams, which also serve helping or rather navigating through the difficulties of calculations.[58] For example, such a diagram may represent the annihilation and then the creation of an electron and a positron via a virtual photon, with one or more virtual photons emitted by an electron later. At any point represented of this diagram yet another virtual process (similar to the emission of virtual photons) may occur and hence another diagram may be inserted into it, thus leading to an interminably expandable rhizomatic structure. Although different events are in principle possible and their possibility defines the situation and what can and cannot *actually* occur, only some of them can be registered. Those particles that are registered by observations are considered as "real particles," while those that are not are considered as "virtual particles." Every virtual or actual event or transition can be represented by a Feynman diagram, and much of quantum field theory consists of drawing and studying such diagrams and generating predictions by using them.

Feynman diagrams are, however, just diagrams, pictures, that help us to heuristically visualize the situation, or, as it were, to *slow down* the phenomenal image of the actual situation of the creation and annihilation of particles, to hold in mind the forms thus created, for the purposes of helping calculations. The same is ultimately true about the "picture" of the creation and annihilation of particles or of the virtual particle formation, or about the particle picture, in the first place. All these "pictures" are useful heuristically or even substantively, in particular in order to handle the multiplicities of actual, registered phenomena, specifically in Bohr's sense, which appears unavoidable under these conditions. What actually happens at the level of such processes themselves we might no more know or even conceive of, let alone visualize, than we can in quantum mechanics, which implies the essential presence of the inconceivable in quantum field theory. Since, in addition, all our knowledge concerning the ultimate constitution of nature is only predictive and, moreover, only statistically predictive,

---

[58] Feynman's *QED: The Strange Theory of Light and Matter* (Feynman 1988) arguably remains the best non-technical book on the subject as well.

chance and, thanks to the predictive capacity of the theory, probability is part of the picture as well, just as they are already in quantum mechanics. In this respect too, however, one encounters a richer play of chance and probability, given the greater multiplicity of phenomena considered. Quantum field theory, again, retains both these epistemological features—the irreducibly inconceivable character of quantum objects and processes, thus disabling the descriptive capacity of the theory, and, correlatively, the irreducibly probabilistic character of the theory's predictions. The corresponding operators or states could be interpreted accordingly in terms of the probabilities of predictions of certain events, analogously (but, again, not identically as far as the procedures used are concerned) to those of the standard quantum mechanics, as discussed in Chapter 2. These conceptions are, however, not sufficient to deal with the situation that quantum field theory has to confront and to build its physical and mathematical architecture. To accomplish this task, it is compelled to engage with the concept of the creation and annihilation of particle and of the virtual particle formation, to become a theory of the birth and disappearance, of quantum objects from the false vacuum, the sea of energy and, epistemologically, the sea of the inconceivable. These new conceptions are or at least at a certain point were necessary in order to develop the mathematical formalism specific to quantum field theory, as against quantum mechanics, such as the creation and annihilation operators, although many (but not all) essential features of the Hilbert space formalism are shared by both theories.

Thus, at least in the present view, this epistemology itself is not different from that found in quantum mechanics as complementarity as concerns what can and cannot be known or ultimately conceived of, or as concerns the essential role of chance and probability. Quantum field theory introduces additional levels complexity and stratification into the situation, especially, again, in terms of richer multiplicities of the events, virtual and actual, involved, which requires significant changes in the formal structure of the theory, including as reflected in Feynman diagrams, a formal structure of a much greater complexity than the formalism of the standard quantum mechanics. As Michael Readhead noted, quantum field theory has more affinity than does quantum mechanics away with the Heracliatean flux and, thus, the Heraclitean Many that with the Parmenidean One (Readhead 1988). Much would depend, however, on how one interprets the Heraclitean flux and the Many, in particular whether one does so in classical or, with Bohr, nonclassical terms. Quantum field theory itself and the Many it brings into play appear to resist classical interpretation just as much or even more than does quantum mechanics. Play is Heraclitus's concept, too, and a very important one for him.

3. "THE ATOMIC STRUCTURE OF THE MEASURING INSTRUMENTS": QUANTUM FIELD THEORY, MEASUREMENT, AND EPISTEMOLOGY

The new (vis-à-vis quantum mechanics) complexities of quantum field theory appear to be deeply connected to the question of measurement, which, as we have seen throughout this study, is so essential to the epistemology of quantum mechanics and to Bohr's interpretation of it as complementarity. Bohr himself addresses the situation already at the early stages of the development of quantum field theory in 1930s. It may be argued that philosophically and epistemologically, the situation is still captured by Bohr's statements made at that time, even though, as I noted, the progress the theory made from quantum electrodynamics, especially following its successful renormalization in the late 1940s, on, culminating in the standard model of fundamental interactions and forces has been immense. I shall cite, first, however, one of Bohr's key statements on measurement in quantum mechanics in "Discussion with Einstein" in 1949. He says: "although, of course, the existence of the quantum of action is ultimately responsible for the properties of the materials of which the measuring instruments are built and on which the functioning of the recording devices depends, this circumstance is not relevant for the problem of adequacy and completeness of the quantum-mechanical description in its aspects here discussed" (*PWNB* 2, p. 51).

As, however, Bohr noted in 1937, in qualifying this point and giving it a different significance, the situation acquires new physical, epistemological, and mathematical complexities, and, with them, Bohr's concept of atomicity (to be developed later) new features, once we move to quantum field theory where the quantum constitution of measuring instruments might need to be taken into account. He says:

> On closer consideration, the present formulation of quantum mechanics in spite of its great fruitfulness would yet seem to be no more than a first step in the necessary generalization of the classical mode of description, justified only by the possibility of disregarding in its domain of application the atomic structure of the measuring instruments themselves in the interpretation of the results of experiment. For a correlation of still deeper laws of nature involving not only the mutual interaction of the so-called elementary constituents of matter but also the stability of their existence, this last assumption can no longer be maintained, as we must be prepared for a more comprehensive generalization of the complementary mode of description which will demand a still more radical renunciation of the usual claims of the so-called visualization. (*PWNB* 4, p. 88)

It may be noted that Bohr's view here can be related to Pauli's argument that in quantum mechanics, as against quantum field theory, the observer is still too "detached" or even "too completely detached" from quantum objects (Pauli 1994, p. 132). "The atomic structure of the measuring instruments" may well be responsible for this detachment and may make it impossible to avoid it. As I noted earlier, given the ultimate form of this epistemology, it is difficult to think of "a still more radical renunciation" than the one complementarity enacts by introducing the irreducibly unthinkable into quantum theory. On the other hand, quantum-field-theoretical phenomena do make us think of (irreducibly) greater forms of multiplicity and hence "new feature of atomicity" (in Bohr's sense) than those found in the case of quantum-mechanical phenomena.[59]

Beyond, self-evidently, Dirac's theory, Bohr's observation just cited must have been prompted in part by Bohr's own work with L. Rosenfeld on measurement in quantum field theory (Bohr and Rosenfeld 1933), in response to L. Landau and R. Peierls's argument, mentioned earlier. The latter argument, contested by Bohr and Rosenfeld, concerned a possible inapplicability of the uncertainty relations in quantum field theory, which is obviously a relevant subject in the present context. This exchange, it is also worth noting, occurs only two years prior to Bohr's exchange with EPR, and, as will be seen below, there manifest affinities between both arguments as regards Bohr's view what can and cannot be unambiguously ascertained in quantum theory, whether quantum mechanics or quantum field theory, given the irreducible role of measuring instruments there. Rosenfeld assisted Bohr in writing his reply to EPR, although one must be cautious in relating their respective views of quantum mechanics and its interpretation, which, I would argue, are quite different. The collaboration, however, clearly support my historical argument that Bohr's work on complementarity also reflects and is shaped by Bohr's thinking concerning quantum field theory, from his introduction by Dirac, while, as noted earlier, working in Bohr's institute in Copenhagen, on. My main concern at the moment, however, is new conceptual and epistemological issues brought in by the question of measurement in quantum field theory in view of the possible role of "the atomic structure of the measuring instruments."

It is important to stress, however, that, as Bohr and Rosenfeld's paper demonstrates (this is indeed its main argument), the adequate (statistical) relationships between the formalism of quantum filed theory and measurements could still be maintained, even though at a further cost (relative to quantum mechanics) as concerns the nature or structure of quantum field theory. As will be seen presently, this assessment

---

[59] It is worth noting that these considerations have deep connections, clearly on Bohr's mind and alluded to by him throughout, to the problem of the physical constitution of the measuring instruments, rods and clocks, in special and then general relativity, which troubled Einstein all his life. Significant though it is in its present context and in its implications (along with the quantum-field-theoretical problematic just discussed) for the deeper understanding of the ultimate constitution of nature (e.g. quantum gravity), I can only mention the subject here, rather than address it. For an illuminating discussion, see (Brown and Pooley 2001).

corresponds to the view of the situation which is still generally accepted and which is still shaped by Bohr and Rosenfeld's paper. According to Bohr and Rosenfeld, here writing on charge measurements:

> It is also of essential importance that the customary description of an electric field in terms of the field components at each space-time point, which characterizes classical field theory and according to which the field should be measurable by means of point charges in the sense of the electron theory, is an idealization which has only restricted applicability in quantum theory. This circumstance finds its proper expression in the quantum-electromagnetic formalism, in which the field quantities are no longer represented by true point functions but by functions of space-time regions, which formally correspond to the average values of the idealized field components over the regions in question. The formalism only allows the derivation of unambiguous predictions about the measurability of such region-functions, and our task thus consists in investigating whether the complementary limitations on the measurability of field quantities, defined in this way, are in accordance with the possibilities of measurement.
>
> Insofar as we can disregard all restrictions arising from the atomistic structure of the measuring instruments, it is actually possible to demonstrate a complete accord in this respect. Besides a thorough investigation of the construction and handling of the test bodies, this demonstration requires, however, consideration of certain new features of the complementary mode of description, which come to light in the discussion of the measurability question, but which were not included in the customary formulation of the indeterminacy principle in connection with non-relativistic quantum mechanics. Not only is it an essential complication of the problem of field measurements that, when comparing field averages over different space-time region, we cannot in an unambiguous way speak about a temporal sequence of the measuring processes; but even the interpretation of individual measurement results requires a still greater caution in the case of field measurements than in the usual quantum-mechanical measurement problem (Bohr and Rosenfeld 1933, *QTM*, pp. 480-481).

It is worth noting that both the trend of thought and the mode of expression of Bohr's reply to EPR is palpably felt here, especially as concern what can and cannot be unambiguous ascertained in quantum theory (quantum mechanics and quantum field theory), which further strengthen my historical argument. For the moment, it is clear that, even though and because they could still be adequately performed, the measurements of quantum fields involves additional levels of idealization, which are the question of "the atomic structure of the field sources and the measuring instrument." Bohr and Rosenfeld's conclusion brings this point into a sharper focus. As they write:

> We thus have arrive at the conclusion, already stated at the beginning, that with respect to the measurability question the quantum theory of fields represents a consistent idealization to the extent that we can disregard all limitations due to the atomic structure of the field source and the measuring instruments. .... This result should properly be regarded as an immediate consequence of the fact that both the [quantum-electromagnetic] formalism and the viewpoints on which the possibilities of testing this formalism are to be assessed have as their common foundation the correspondence argument. Nevertheless, it would seem that the somewhat complicated character of the considerations used to demonstrate the agreement between formalism and measurement are hardly avoidable. For in the first place the physical requirements to be imposed on the measuring arrangement are conditioned by the integral form in which the assertions of the quantum-electromagnetic formalism are expressed, whereby the peculiar simplicity of the classical field theory as a purely differential theory is lost. Furthermore, as we have seen, the interpretation of the measuring results and their utilization by means of the formalism require consideration of certain features of the complementary mode of description which do not appear in the measurement problems of non-relativistic quantum mechanics. (Bohr and Rosenfeld 1933, *QTM*, pp. 520-521)

These complexities appear to relate essentially to the complexities of quantum field theory that are still with us. This assessment is supported by the view of some of the founders of modern (i.e., renormalized) quantum electrodynamics, in particular, Julian Schwinger and Freeman Dyson, consistent with that of Bohr or Pauli, as here considered. According to this view the appearance of the infinities or divergencies and the necessity of renormalization (on which I shall comment presently) arise in view of the quantum-mechanical, as against, quantum-field-theoretical, idealization of measurement of the type invoked by Bohr. Indeed, Dyson addressed the situation, via, on the one hand, Heisenberg, and, on the other, Bohr and Rosenfeld's analysis of measurement in quantum field theory, by virtually repeating Bohr's commentary just cited, but giving it a properly modern connection to the formalism of quantum field theory. He writes:

> We interpret the contrast between the divergent Hamiltonian formalism [which imposes the necessity of renormalization] and the finite *S*-matrix as a contrast between two pictures of the world, seen by two observers having a different choice of measuring equipment at their disposal. The first picture is a collection of quantized fields with localizable interactions, and is seen by a fictitious

> observer whose apparatus has no atomic structure and whose measurements are limited in accuracy only by the existence of the fundamental constants c and h. This ["ideal"] observer is able to make with complete freedom on a sub-microscopic scale the kind of observations which Bohr and Rosenfeld employ ... in their classical discussion of the measurability of field-quantities. The second picture is of collection of observable quantities (in the terminology of Heisenberg) and is the picture seen by a real observer, whose apparatus consist of atoms and elementary particles and whose measurements are limited in accuracy not only by c and h, but also by other constants such as α [the fine-structure constant] and m [the mass of the electron]. (Dyson 1949, p. 1755, cited in Schweber 1994, p. 548)

Silvan Schweber, who cites the passage, comments on it as follows, further clarifying the situation from the modern, post-renormalization, quantum-field-theoretical perspective:

> A "real observer" can measure energy levels and, and perform experiments involving the scattering of various elementary particles—the observables of *S*-matrix theory—but cannot measure field strengths in small regions of spacetime. The "ideal" observer, making use of the kind of "ideal" apparatus described by Bohr and Rosenfeld, can make measurements of this last kind, and the commutation relations of the fields can be interpreted in terms of such measurements [reflected in the corresponding quantum-field-theoretical uncertainty relations]. The Hamiltonian density will presumably always remain unobservable to the real observer whereas the ideal observer, "using nonatomic apparatus whose location in space and time is known with infinite precision" is presumed to be able to measure the interaction Hamiltonian density. "In conformity with the Heisenberg uncertainty principle, it can perhaps be considered a physical consequence of the infinitely precise knowledge of location allowed to the ideal observer that the value obtained by him when he measures Hamiltonian density is infinity" (Dyson 1949, p. 1755). If this analysis is correct, Dyson speculated, the divergences of QED are directly attributable "to the fact that the Hamiltonian formalism is based upon an idealized conceptual of measurability" (p. 1755). (Schweber 1994, p. 548)

This is quite possible, since, as we have seen, the Hamiltonian formalism was brought into quantum theory, first, via the correspondence principle into quantum mechanics by Heisenberg, and then transferred by Dirac into quantum electrodynamics, where the correspondence considerations of the type used by Heisenberg are no longer applicable, as Bohr suspected immediately (*PWNB* 4, p. 56). But there may be no better formalism. I am, of course, not saying that a better formalism is not possible mathematically. It is possible and may become necessary, even leaving aside how far the present-day formalism traveled from the original one, although perhaps not yet far enough. My point instead is that no formalism may be able to overcome the physical and epistemological constraints of the situation. Or, one might say, it cannot overcome a certain enclosure of it, which cannot be transcended, physically and conceptually, even though that which is beyond this enclosure shapes and reshapes via certain effects, the peculiar effects, defining quantum phenomena, those in question in quantum mechanics and those in question in quantum field theory.

In other words, the "ideal" observer is in fact the quantum-mechanical observer, which is to say, as Bohr very likely would, first, that it is the only real observer that is actually possible, given the measuring instruments we have, the perceptual machinery of our body (and the conceptual machinery of our mind) included. The idealization in question is clearly at stake in Bohr and Rosenfeld's discussion of measurement in quantum field theory, as considered above. We observe the quantum-field-theoretical effects via or as mediated by quantum-mechanical effects.[60] Since, on the present view, each of these effects is actually observed or measured as a classical physical objects, this real observer is in fact a classical observer, but (this is of course crucial) as a classical observer of quantum-mechanical effects. As discussed earlier, Heisenberg's uncertainty relations in fact presuppose the existence of the infinitely precise instruments. In this sense, this observer is indeed ideal, and the actual reality of the quantum-mechanical, or the quantum-field theoretical, observation brings with it the statistical considerations and, ultimately the irreducibly statistical character of all quantum theories. I leave aside for the moment the idealization involved in classical physics, since it does not affect my argument here, and specifically the point that the "real" observer of Dyson is not realizable, in principle, not, one might say, *humanly* possible.

In short, the epistemology in question is precisely that of Bohr's complementarity, as considered in this study. Quantum field theory is a new set of, in Bohr words, "mathematical instruments," coupled to new measuring instruments and new types of the configurations of phenomena or "atoms" in Bohr's sense. As such, quantum field theory allow us to handle the unknowable or even the unthinkable nature of quantum objects and processes, that manifest its existence, although, again, not themselves, in new type of effects we observe at the higher energy level.

Julian Schwinger, whose philosophy of quantum physics, may be argued to be close to that of Bohr, makes this point in the context of renormalization. I cannot address

---

[60] Bohr and Rosenfeld's collaboration already orients our thinking in this direction. I am indebted to Giuseppe Vitiello for the discussion of this subject. For some intriguing ramifications of this point and of the difference between quantum mechanics and quantum field theory via some powerful and elegant mathematics, such as Hopf's algebras, see (Iorio, Lambiase, and Vitiello 2002).

the subject of renormalization in detail here, and can only give a brief summary of what is at stake courtesy to Dyson, which is necessary to appreciate Schwinger's point.[61] According to Dyson:

> [Quantum fluctuations of the electromagnetic field in the atom, say, the hydrogen atom] would give the electron an additional energy, called the self-energy. It was well known that [Dirac's] quantum electrodynamics (QED) gave an infinite value for the self-energy and was therefore useless. Physics has reached an impasse. On the one hand, the Lamb experiment gave clear evidence that the effects of electromagnetic quantum fluctuations were real and finite. On the other hand, the existing theory of QED gave infinite and absurd results. . . .
>
> [In 1940s, however, Kramers] remarked that the observed energy of an electron , according to QED is the sum of two unobservable quantities: a bare energy, which electron is supposed to have when it is uncoupled from electromagnetic fields, and the self-energy, which results form the electromagnetic coupling. The bare energy appears in the equations of the theory but is physically meaningless, since the electromagnetic coupling cannot really be switched off. Only the observed energy is physically meaningful. The point of renormalization was to get rid of bare energies and replacing them with observed energies. (Dyson 2005, p. 48)

This was is what was essentially accomplished, thus restoring the legitimacy to Dirac's theory, rather than abandoning it altogether, as some, Dirac among them, proposed, and replacing it with a different theory. It is clear that the question of measurement may have essential bearings on the situation, along the lines just discussed, although the questions thus posed remains formidable and seemingly out of reach. Accordingly, Bohr's radical epistemology (the irreducibly inaccessible nature of quantum objects and processes), the quantum-field-theoretical multiplicity of phenomena (conceptually linked to the creation and annihilation of quantum objects and to the virtual particle formation), and the role of the atomic constitution of the measuring instruments becomes all connected. One might, accordingly, say that, at least potentially, these connections define what might be called the epistemological problem of quantum field theory. This is a problem of great difficulty, especially, if one wants to link to rigorously to the mathematical formalism of the theory, at least to the same degree as in

---

[61] For a lucid and (reasonably) accessible account, see (Teller, pp 149-168) and for a more historically oriented account see the relevant discussion in (Schweber 1994), especially "Epilogue" (pp. 595-605). More recent developments, such as renormalization group, effective quantum field theory, and so forth, cannot be pursued here, in part because of their nearly prohibitive technical aspects. On the other hand, they also do not appear to me to change the epistemological argument here offered.

quantum mechanics, and as such it is bound both to lead us beyond Bohr and make it difficult to leave him behind.

The renormalization procedure is extremely difficult mathematically, and its mathematical legitimacy is as yet an unresolved issue.[62] In the case of quantum electrodynamics, it was effectively performed in 1940s by Tomonaga, Schwinger, and Feynman, which brought them a joint Nobel Prize in 1965, with some contribution by others, especially Dyson, or earlier Hans Bethe and Hendrik Kramers. The Young-Mills theory was eventually shown to be renormalizable as well, by Martinus Veltman and Gerardus t'Hooft in 1970s (eventually bringing them their Nobel prize as well). This allowed a proper development of the standard model of all forces of nature, except for gravity, which is, as yet, not given its quantum form. Now, according to Schwinger:

> [Renormalization] is the clear separation of what we don't know—but which affects our experiments in a very limited way—from what we do know and where we can calculate in detail. In fact, I insist that all theories are like this. — People may not want to face up to it, [but] there is always an area beyond where the theory either breaks down or where other phenomena come into play that you don't know about. They do not upset everything in the area you can control, and you isolate that from it: That's what renormalization is really about. Not sweeping infinities away but isolating the unknown part and recognizing its limited influence. (Schwinger 1982 [taped interview with Schweber]; cited in Schweber 1994, p. 366)

This type of view or approach, "isolating the unknown," is anticipated by Dirac at the early stages of the history of this question, as is indeed indicated in his letter to Bohr, some time in August 1933 (Dirac 1933), and these ideas, I argue here, had a significant impact on Bohr's view of quantum field theory.[63] In the case of quantum field theory, however, or already in quantum mechanics, it may not be possible to move beyond the limits that we were able to isolate thus far. In this respect, contrary to Schwinger's claim, quantum theory may not be like any other theory hitherto

---

[62] Roughly speaking, the procedure might be seen as manipulating infinite integrals that are divergent and, hence, mathematically illegitimate. At a certain stage of calculations, these integrals are replaced by finite integrals through artificial cut-offs that have no proper mathematical justification and are performed by putting in, by hand, experimentally obtained numbers that make these integrals finite, which removes the infinities from the final results of calculations. See, again, Teller's discussion in (Teller, pp 149-168). These calculations are experimentally confirmed to a very great degree. Indeed, quantum electrodynamics is the best experimentally confirmed theory in our possession.

[63] See Pais's discussion of this important letter in (Pais 1986, pp. 382-383).

encountered. In quantum field theory, the effects of the unknowable or the unthinkable appear to require a more complex mathematical machinery and new techniques such as renormalization, because of the infinities, while retaining this unknowable or unthinkable as part of our knowledge and thinking, even as their enabling part, which is Bohr's point concerning quantum field theory, as here discussed. As to the question whether there could be some underlying finite theory, for example, a form of string or brane theory, that, it is hoped, will eventually enable us to avoid the infinities of quantum field theory, handled via renormalization, one might be best off by adopting, at least for now, Pais's "answer," cited above: "Perhaps, perhaps not." The same answer could be given in response to the question whether such a theory, or any future theory, will allow us to avoid the epistemological complexities here discussed. Quantum mechanics is finite, but it led us to these complexities.

The epistemology of quantum field theory and of the phenomena in question in it appears to nearly inevitably lead to great and perhaps as yet unperceived and hitherto unimagined or unimaginable complexities, although, as I said, epistemologically it is difficult to move beyond the quantum-mechanical renunciation, introduced by complementarity, concerning what can possibly be known and thought of. These complexities are likely to move us even further beyond Bohr. We cannot, however, be altogether certain in what direction. Either way, it might be difficult to leave Bohr behind, but we continue to discover new ways of reading Bohr.

Chapter 6. Complementarity: From Physics to Philosophy, From Philosophy to Physics

1. INTRODUCTION: THOUGHT, KNOWLEDGE, AND CONCEPTS IN PHYSICS AND PHILOSOPHY

This chapter offers a *philosophical* discussion of the epistemology and conceptuality arising from Bohr's interpretation of quantum mechanics as complementarity, or from the interpretation of this interpretation offered in this study. Following the terminology adopted in Chapter 4 in the context of quantum probability, I shall call this epistemology "nonclassical." As discussed earlier in this study, Bohr's adjustments of complementarity were sufficiently significant to allow one to speak of several versions of it. Only one of these versions, that developed in the wake of EPR's argument and finalized sometime in the 1940s, and defined by Bohr's concepts of phenomenon and then atomicity, appears to fully conform to nonclassical epistemology. It is, in principle, possible to speak of several post-EPR versions of complementarity as well. The possibility of considering several such versions, whether all of them are nonclassical or not, would not, however, affect my argument here. That the one found in "Discussion with Einstein" (1949) and "Quantum Physics and Philosophy: Causality and Complementarity" (1958) is nonclassical, at least in the present interpretation, suffices for an argument, such as the one to be offered here, concerning both the possibility and significance of nonclassical epistemology in physics as based on actually existing theories or interpretations of these theories. In addition, as discussed in Chapter 5, quantum field theory appears, at the very least, to allow for interpretations that are epistemologically nonclassical.

Obviously, epistemology alone is not sufficient for a project such as that of complementarity, since this project requires an interpretation of both quantum phenomena themselves and quantum mechanics, and, hence, concepts and language (physical, mathematical, philosophical, or ordinary), old and new, through which and only through which such an interpretation is possible. One can, accordingly, also speak, as I shall do here, of nonclassical *theories*—that is, theories, such as complementarity (comprised by quantum phenomena, quantum mechanics, and Bohr's particular interpretation of both), that conform to nonclassical epistemology but are not restricted to it. To the degree that old concepts, such as those of classical physics in quantum mechanics, are used by a nonclassical theory, they must be made physically and epistemologically consistent with this interpretation. Accordingly, such a use requires a critique of these concepts in the Kantian sense of an analysis of their architecture in order to establish the limits within which they, or some of their components, can be used in a new regime. For example, in Bohr's interpretation, the key concepts of classical physics still apply rigorously to measuring instruments (under the constraints of the uncertainty relations) and only in a provisional or symbolic way to quantum objects. Bohr's new concepts—such as those of complementarity (in the narrow sense of mutual exclusivity of certain entities), phenomenon, and atomicity, and his concept of quantum objects and processes as beyond any conceptualization—incorporate and are in part based on this epistemology. It is by coupling these concepts to this epistemology that Bohr is able to offer a consistent interpretation of quantum mechanics as, within its scope, a complete and local theory of quantum phenomena.

Accordingly, Bohr's philosophy of quantum mechanics is defined by three fundamental features—first, a new epistemology of quantum phenomena and quantum mechanics; second, a critique of already available physical and philosophical concepts, such as those of classical physics; and third, the invention of new concepts. The aim of this chapter is an exploration of these features and of the relationships among them, and of a broader philosophical problematic that emerges in view of this understanding of Bohr's thought.

As this book as a whole, this chapter aims to delineate the general philosophical content of Bohr's epistemology and his key concepts as relatively independent of, but not altogether divorced from, their physical content. A "removal" of physics from the philosophical architecture of Bohr's concepts to use them independently of their physical content elsewhere is possible, as Bohr indicated on several occasions in his philosophical essays, and as was noted (and sometimes done) by others, including previous occasions by the present author (Plotnitsky 2004). This *possibility* is important for my argument in this chapter, in particular insofar as it supports and amplifies the view that these concepts also have a philosophical, rather than only physical, content and that this philosophical content is viable in its own right. I shall not, however, pursue my argument through a subtraction, or abstraction, of Bohr's concepts from their physical content, even though this argument concerns the *philosophical* nature and impact of Bohr's thought, or of the thought of Heisenberg, which I shall also address in this chapter. There are several reasons for adopting this approach, an approach that combines physics and philosophy in order to explore philosophy.

First of all, Bohr's physics and indeed physics, to begin with, allow one to elucidate and to explore the philosophical content of these concepts and nonclassical epistemology itself in greater depth and with greater effectiveness. Their exposition is difficult enough even when helped by the concreteness of physics; a purely abstract exposition is, again, possible but is not likely to be desirable by most readers, especially (but not exclusively) physicists.

Secondly and more importantly, as I have argued in this study, as an interpretation of quantum mechanics, Bohr's complementarity is primarily physics, substantively and disciplinarily, and its more generally philosophical dimensions primarily serve its physics. I would contend this to be the case even though, as just indicated, Bohr was not adverse to more independent or of course interactive philosophical developments of his concepts and noted the relationships between complementarity and epistemological or conceptual developments in other fields throughout his writings. Indeed, he sometimes invoked his "dream of great interconnections" between different fields of thought or human endeavors. Also, as is well known, certain key aspects complementarity (in particular the idea that certain mutually exclusive phenomena can be used within the same theoretical framework) originate in Bohr's earlier interests in philosophical psychology, the field he initially considered pursuing professionally. This idea comes in part from Bernhard Riemann's approach to resolving the problem of the two-valued-ness of certain functions of complex variables, which led him to the idea of Riemann surfaces, one of his several great contributions to modern mathematics. In general, the conceptual architecture of complementarity has a complex genealogy, which at some point led Bohr to coupling certain ideas that came from outside physics (e.g., philosophy, psychology, or mathematics) to physics with this conceptual architecture as an outcome. On the other hand, while Bohr, again, often invoked the possibility of doing so, he never pursued the relationships between complementarity and other fields (elsewhere in science, in psychology, or in philosophy) beyond speculative and tentative suggestions, or developed complementarity accordingly. Some commentators give these suggestions or these connections a greater weight than I am inclined to do here or than, in my view, Bohr gave them himself, although he, again, stressed the potential significance of such connections. In any event, I would contend that it is the physics of complementarity that represents best and is primarily responsible for its philosophy.

Accordingly, I argue that it is through this physics, rather than Bohr's more overtly philosophical speculations, that we understand the philosophy of complementarity best and can develop it best, possibly beyond Bohr, without, again, leaving him behind. In other words, in parallel with the movement, sketched earlier in this study, beyond Bohr without leaving him behind (which movement involves both physics or philosophy), one can, as I shall attempt here, move beyond physics to philosophy without leaving physics behind, or in the opposite or reciprocal direction beyond philosophy to physics without leaving philosophy behind.

Bohr himself makes this type of point, thereby also placing his own work in the history of physics' contribution to philosophical thinking, in the opening sentence of his "Introduction" to the second collection of his philosophical essays, *Atomic Theory and Human Knowledge*, originally published in 1958 (now Volume 2 of *PWNB*). He says: "The importance of physical sciencefor the development of general philosophical thinking rests not only on its contribution to our steadily increasing knowledge of that nature of which we ourselves are part, but also on the opportunities which time and again it has offered for examination and refinement of our conceptual tools. In our century, the study of the atomic constitution of matter has revealed an unsuspected limitation of the scope of classical physical ideas and has thrown new light on the demands on scientific explanation incorporated in traditional philosophy" (*PWNB* 2, p. 1; see also *PWNB* 3, p. 1). An exploration (again, beyond physics or Bohr but without leaving them behind) of some among the opportunities offered by complementarity or by quantum theory itself, both quantum mechanics and quantum field theory, is the aim of this chapter.

According to Heisenberg, "Bohr was primarily a philosopher, not a physicist, but he understood that *natural* philosophy in our day and age carries weight only if its every detail can be subjected to the inexorable test of experiment" (Heisenberg 1967, p. 95; emphasis added). While it is difficult to contest Heisenberg's contention that natural philosophy, that is, a philosophy of nature, in our day and age carries weight only "if its every detail can be subjected to the inexorable test of experiment," one might question whether Bohr was primarily a philosopher, rather than a physicist, even in the case of his work on complementarity. Heisenberg's comment, made around the time of Heisenberg's first stay in Copenhagen, refers to his encounter with Bohr well before the introduction of complementarity or even quantum mechanics itself, but is, I think, aimed to describe Bohr's thinking in general. I would argue Bohr was both, a physicist and a philosopher, and perhaps ultimately more a physicist than a philosopher. A physicist as a philosopher? A philosopher as a physicist? One's view of his thought and work may ultimately depend on one's definition of either field or form of thought, physics and philosophy. In any event, I would argue that he was at his best philosophically when dealing with physics. Heisenberg's remark, however, rightly conveys the fundamental significance of the relationships between physics and philosophy in Bohr, or, as will be seen, in Heisenberg's own work, perhaps following Bohr, although it was ultimately more mathematically and less philosophically oriented (Heisenberg 1967, p. 95).

Accordingly, as I noted above, the physical and the philosophical content of Bohr's concepts mutually symmetrically define and elucidate each other. In the process they help us to understand better the reciprocity between physics and philosophy in the architecture of these concepts and in Bohr's view of quantum phenomena and quantum mechanics. This reciprocity remains as crucial to my argument in this chapter as it has been throughout this study. The difference is that of a relative emphasis, which this chapter places on the philosophical dimensions of Bohr's concepts and thought. The discussion of Heisenberg's thinking in general and specifically his critique of classical physical concepts as a form of Kantian critique in Section 4 of this chapter in part serve to illustrate this reciprocity as well. Indeed, as the title of this chapter indicates, my goal here is also to explore how this reciprocity allows us to move in both directions—from physics to philosophy and from philosophy to physics. Both types of movement are found throughout Bohr's work on complementarity and in Heisenberg's work to be discussed here.

I shall, however, argue that the nonclassical character of thinking and knowledge concerning quantum phenomena and quantum mechanics entails and, along with certain developments elsewhere, historically led to a transformation of our ideas concerning the nature of all our thinking and knowledge, especially as concerns our construction of theories in physics or elsewhere. Bohr's title for the collection just cited, *Atomic Theory and Human Knowledge*, directs us to these connections as well. In his introduction, which I cited above, Bohr announces the epistemological argument pursued in the essays assembled there as an argument that concerns both quantum mechanics itself and "the conditions for the proper use of our conceptual means of expression" in general. He says:

> The importance of physical science for the development of general philosophical thinking rests not only on its contributions to our steadily increasing knowledge of that nature of which we ourselves are part, but also on the opportunities which time and again it has offered for examination and refinement of our conceptual tools. [...] The main point of the lesson given us by the development of atomic physics is, as is well known, the recognition of wholeness in atomic processes, disclosed by the discovery of the quantum of action. The following articles present the essential aspects of the situation in quantum physics itself and, at the same time, stress the point of similarity it exhibits to our positions in other fields of knowledge beyond the scope of the mechanical conception of nature. We are not dealing here with more or less vague analogies, but with an investigation of the conditions for the proper use of our conceptual means of expression. (*PWNB* 2, pp. 1-2)

"The mechanical conception of nature" invoked by Bohr here is also defined by the classical, Democritean view of atomicity, replaced by Bohr's concept of atomicity, via his concept of phenomena, as discussed earlier, a concept that reflects "the essential aspects of the situation in quantum physics." Bohr links this "novel situation in physical science" to "clarifying the condition of objective description in wider fields," which may and in some cases do conform to the same (on the present definition, nonclassical) type of epistemology. Although in general a more complex term in Bohr, "objective" primarily means here unambiguously communicable, as is made clear and directly stated throughout his later works. As he says in a later essay, "Quantum Physics and Philosophy: Causality and Complementarity" (1958): "The description of atomic phenomena has in these respects [i.e. as defined by his concept of phenomena] a perfectly objective character, in the sense that no explicit reference is made to any individual observer and that therefore, with proper regard to relativistic exigencies, no ambiguity is involved in the communication of information" (*PWNB* 3, p. 3). Bohr's appeal to a "proper regard to relativistic exigencies" is worth registering here as yet another testimony that he views quantum mechanics as complementarity as a local theory. My main point at the moment, however, is that, as nonclassical epistemology, the epistemology in question fundamentally or, to use again, the term favored by Bohr, *essentially* transforms the nature of our thinking and knowledge and the conditions of the proper use of our fundamental concepts, including the very concepts of thinking and of knowledge. Accordingly, this transformation also makes a major contribution to and in turn transforms our philosophical thinking, since the nature of thinking and knowledge is one of the primary concerns of philosophy.

As noted above, such concepts may be old, such as, in quantum mechanics, those inherited from classical physics, and new, such as those developed in quantum mechanics or those specifically introduced by Bohr, for example, complementarity, phenomena, and atomicity. Bohr's own use of old concepts or his invention and deployment of new concepts bring together nonclassical epistemology and conceptual thinking. The use of old concepts would now entail a *critique* of these concepts in the Kantian sense of deeper understanding of their nature and workings even in classical physics and of properly establishing the limits of the applicability of these concepts or some of their components in a new (nonclassical) epistemological regime. This type of critique is undertaken by Bohr throughout his work (as it is, as I shall discuss, by Heisenberg in his Chicago lectures, *The Physical Principles of the Quantum Theory* [Heisenberg 1930]). The significance and even presence of this critique is, as we have seen, often missed by commentators on Bohr's argument for the indispensability of classical concepts in quantum mechanics. In particular, according to Bohr, first, the physical application of classical physical concepts (i.e., those that define classical physics) is rigorously possible only in considering phenomena and processes in which Planck's quantum of action, *h*, could be neglected, which, accordingly, allows for their rigorous application in classical physics. (As explained earlier in this study and as will be further discussed below, via Heisenberg, relativity already poses certain difficulties in using some of these concepts and entails their critique.) Secondly, this rigorous applicability of classical physical concepts within the proper limits of classical physics allows one to apply them to the physical behavior of certain parts of measuring instruments in considering quantum phenomena. We recall, however, that, as against the phenomena in question in classical physics, this is now only possible under the constraints of the uncertainty relations. The latter in Bohr's interpretation strictly apply to these parts of measuring instruments, while certain other parts (those enabling their interactions with quantum objects) of these instruments are seen as quantum in their constitution. Finally, classical physical concepts are seen by Bohr, or by Heisenberg, as a refinement of our everyday perception and thinking, which, however further refined, may not be able to access the actual nature or, assuming, again, we can apply even this term, "reality" of quantum objects and processes.

As will be explained in detail below, the very term classical could be extended accordingly—that is, to what can be in principle represented or thought of. In this sense, all concepts we can form are classical. Under these conditions, quantum mechanics is "saved" by the mathematics of quantum theory. This mathematics allows one to make correct, albeit generally only statistical, predictions concerning the phenomena in question as outcomes of certain rigorously specifiable experiments. That it does so is inexplicable or enigmatic, given that it does not describe, even in an idealized fashion, the physical processes involved.

This argument positions Bohr's thought, and that of Heisenberg, between Kant's and Hegel's philosophical thought. The confrontation between Kant and Hegel (which, it appears, often brings Hegel closer to Kant, in spite of his attempt to move away from him) has defined the history of modern philosophy from Hegel on and still overshadows philosophical thought. As will be seen, this is also true as concerns the Bohr-Einstein debate on the epistemology of quantum mechanics. Both Einstein and Bohr, again, occupy a complex position between Kant and Hegel, and I shall comment on Einstein's relationships to both below. Thus, both of the relationships that crucially shaped Bohr's thought on quantum physics, that between Bohr and Heisenberg (defined primarily by affinities) and between Bohr and Einstein (defined primarily by differences and confrontation) may be seen through the optics of the confrontation between Kant and Hegel.

Bohr follows, but also radicalizes, Kant along the lines of epistemology and Hegel as concerns the significance of new concepts. He follows equally both as concerns the critique of old concepts and the conditions of the deployment of new concepts, or, in Kant's phrase, defining his critical or, as it also called transcendental (not the same as transcendent) philosophy, the conditions of the possibility of concepts, objects, or phenomena. The confrontation itself between Kant and Hegel is defined by the fundamental relationships, found in both thinkers but with a different balance in each case, between the indispensability and power of concepts and yet their potential and potentially irreducible insufficiency in grasping or even approaching the world, natural or human. The question of these relationships, I argue, finds its ultimate manifestation, at least thus far, in nonclassical epistemology, such as that of Bohr. There are other thinkers who could be mentioned here, such as and in particular Friedrich Nietzsche, who offered a more radical critique of the concepts of reality and causality than anyone before him and, I would argue, as radical as anyone since, including Bohr. Nietzsche's ideas, it is worth noting, were known to Bohr, through Georg Brandes, a philosophy professor at the University of Copenhagen, who gave the first ever university course on Nietzsche. Brandes was a friend of Bohr's father and was greatly admired by Bohr. Nietzsche's thought, too, or those of his key followers in the twentieth century, places itself in relation to the confrontation between Kant and Hegel, although, like that of Bohr, it moves into new territories of epistemology and conceptuality. It is always, in Nietzsche's famous phrase, "a philosophy of the future."

My argument, however, is not primarily concerned with tracing Bohr's epistemology and his concepts, such as those of complementarity, phenomena, or atomicity, to those of Kant and Hegel, or other philosophical figures, such as William

James or Harald Høffding, often mentioned in commentaries on Bohr, including by the present author (e.g., Faye 1991, Folse 1985, Mittelstaedt 1994, and Plotnitsky 1994). To some degree, such a tracing is unavoidable and will, accordingly, be pursued here as well, but only as an auxiliary part of my discussion. Nor is my argument aimed at placing Bohr's thought in the context of and in relation to the more technically oriented academic philosophy of quantum mechanics. In addition to giving the main emphasis to the mathematical formalism of quantum mechanics, the practitioners in this field by and large follow the tradition of Anglo-American analytic philosophy, rather than that of continental philosophy, extending primarily from Kant and Hegel, or in the last century Husserl and Heidegger. This type of placement of Bohr's thought could be an interesting project, and the links between approaches of that type and, at least, the spirit of Copenhagen have not been altogether absent (e.g., van Fraassen 1991). My argument may be seen not so much in contrast to either type of projects as complementing them (in this case not in Bohr's sense of complementarity as the mutual exclusivity of certain entities). My primary aim, however, is the more fundamental significance of Bohr's, or Heisenberg's, epistemology and conceptual architecture in contemporary thought in physics and philosophy, as defined, as I would argue it still is, by the great confrontation between Kant and Hegel.

Through this program, Bohr's philosophy or, by implication, the philosophy of quantum mechanics is given a broader sense defined by the three fundamental features noted above. The first is epistemology in the broad sense of rethinking the nature of knowledge in physics and beyond; the second is a (Kantian) critique of already available physical and philosophical concepts, such as those of classical physics (which are also philosophical); and, finally, the third is the invention of new concepts. The third aspect of philosophical thinking just stated, the invention of new concepts is crucial to Bohr, although it is often missed or misunderstood in the commentaries on Bohr. This misunderstanding is in part due to an even more common misunderstanding, noted above, of Bohr's insistence on the continuing significance of the classical physical and philosophical concepts in quantum theory. I have considered the subject throughout this study and shall further comment on it later in this chapter. It is, however, worth reiterating at this point that this part of Bohr's argument is best understood in terms of a (Kantian) *critique* of classical concepts and in part a redefinition of what classical concepts are, rather than (as some of the commentaries just mentioned see it) as an uncritical transplanting of concepts from classical to quantum physics. On the other hand, as a practice of invention and construction, of building up new concepts, Bohr's philosophy of complementarity may be best seen in accordance with Gilles Deleuze and Félix Guattari's view of philosophy itself as the creation of new concepts, a view that is in turn close to that of Hegel (Deleuze and Guattari 1993, p. 5).

According to Deleuze and Guattari, the creation of new concepts defines philosophy and is the essential task of philosophy's *thought*. As such, it is also the way in which this thought confronts chaos, a confrontation that, according to them, defines thought qua thought, wherever it operates, in philosophy, science, or art. I shall return to this argument in closing this chapter. This view of philosophy as the creation of new concepts also implies a different view or a different concept of concept itself, concomitantly introduced by Deleuze and Guattari, who, again, follow the post-Kantians and especially Hegel. Hegel defines philosophy through the primary role of the concept (*das Begriff*) in the same sense or concept of concept (Deleuze and Guattari 1993, pp. 11-12). In this view, a philosophical concept is not, as conventionally understood, an entity established by a generalization from particulars (e.g., the concept of tree, derived by generalizing the images of all different individual trees into a single conceptual entity) or any general or abstract idea. Instead, it is a complex, multi-layered structure or architecture—a multi-component conglomerate of concepts in their conventional sense, images and figures, particular elements, and so forth—and as such may define a whole philosophical matrix. Bohr's concepts, such as complementarity, phenomenon, atomicity, or quantum objects, are just such concepts. We recall that, according to Frank Wilczek, "the primary goal of fundamental physics is to discover profound [physical] concepts that illuminate our understanding of nature" (Wilczek 2005, p. 239). Bohr's concepts were certainly developed by Bohr as physical concepts and in order to pursue this goal. They may, however, also be seen as philosophical concepts and, as explained above, were seen by Bohr as such concepts; and, I argue, that reciprocally with the picture just sketched of the philosophical architecture of these concepts as containing a physical component, their architecture as physical concepts contains an irreducible philosophical component. The greatness of their philosophical architecture, I also argue, comes from this reciprocity between physics and philosophy.

Bohr's writings on quantum mechanics, or earlier writings by Galileo, may demonstrate this reciprocity especially powerfully. The work of most major figures in the history of physics, however, certainly that of Einstein, would be nearly as indicative here, or, conversely, the work of major philosophical figures, such as, and in particular, Kant or Nietzsche. Nietzsche indeed once said: "Und darum: Hoch die Physik! Und höher noch das, was uns zu ihr zwingt,—unsre Redlichkeit!" [And this is why: long live physics! And even more so that which forces us to turn to it—our integrity!] (Nietzsche 1974, p. 266; translation modified). Nietzsche's remark may be seen as part of his more general argument for putting everything we claim—in science, philosophy, or ethics—to "the inexorable test of experiment," to return to Heisenberg's description of Bohr's understanding of any philosophy of nature "in our day and age." "To that end," Nietzsche says, "we must become the best learners and discoverers of everything that is lawful and necessary in the world: we must become *physicists* is order to be able to be creators in this sense" (Nietzsche 1974, p. 266). Nietzsche most likely had in mind such contemporary developments as electromagnetism and, especially, thermodynamics, which, at the time, put contemporary physical and philosophical conceptions to a severe test indeed. Twentieth- and by now twenty-first century physics, and much of its mathematics and science in general, have continued to test our most fundamental philosophical concepts, all our concepts, and to force us to develop new ones.

In the same book, Nietzsche offers a parallel view of mathematics and its role in all sciences or, it appears, philosophy, with a Kantian and even nonclassical touch, as something that helps us to "*determine* our human relations to things," rather than "to know things." He says: "*Mathematics.*— Let us introduce the refinement and rigor of mathematics into all sciences as far as this is at all possible, not in the faith that this will lead us to know things but in order to *determine* our human relations to things. Mathematics is merely the means for the general and ultimate knowledge of man" (Nietzsche 1974, p. 215). That is, it is the means for the general and ultimate knowledge of how we think and know, and Nietzsche is likely to allude to and to (re)interpret Protagoras' famous statement that "man is the measure of all things." Merely, but crucially! Nietzsche's German for science is *Wissenschaft*, which thus would also include philosophy or psychology as well, where we, thus, must also use mathematics or mathematical ways of thinking, "as far as this is at all possible"—a wise qualification. With Heisenberg, the mathematics of quantum mechanics no longer even allows one to know things, the ultimate constitutions of things, but only to predict certain things. Well, not quite! For, in accordance with Nietzsche's maxim, it enables us to know that there are certain things that we cannot know or cannot even conceive, and hence, again, to maintain our integrity as thinkers, learners, and knowers. "And this is why: Long live physics!"

Reciprocally, however, nonclassical philosophy may help physics and sometimes compel physics to turn to philosophy as part of physics' integrity, which, accordingly, becomes defined by philosophy, along with experiment and mathematics. In other words, both have reciprocally tested and advanced, and created each other throughout their history, and, in the process, have productively shaped the ambient culture around them, and continue to do so.

2. NONCLASSICAL EPISTEMOLOGY AND ITS CONCEPTS

I would like to offer first a general definition of a *nonclassical* theory, as a theory conforming to nonclassical epistemology. This definition is paradigmatically established by quantum mechanics as complementarity, as considered and interpreted in this study, but is applicable beyond it. In the case of quantum mechanics, the denomination nonclassical theory would, *in this context*, apply to a given nonclassical interpretation of quantum mechanics, that is, a physical and epistemological interpretation of the mathematical formalism cum the experimental data or phenomena in question in quantum mechanics. In general, however, a nonclassical theory *defines* certain (one may call them "nonclassical" in turn) objects and processes as being beyond the reach of the theory itself, or even of any knowledge or conception available to us. They are unthinkable in the literal sense of *un*-thinkable as being *beyond thinking*, even as "objects" or "processes" in any conceivable sense (quotation marks become, accordingly, necessary). Crucially, however, such a theory considers these objects and processes as essentially responsible for what it can observe, know, and think, and what it must account for. Indeed, such objects and processes are rigorously defined by means of, or *from within*, this theory and are, accordingly, an *essential* concern, or even the most essential concern, of the nonclassical theory that introduces them, even though and because the theory is unable to say anything about or conceive, *think*, of them and even is precluded from doing so. Thus, Bohr's complementarity, as a nonclassical interpretation of quantum phenomena and quantum mechanics, does not imply (along more positivist lines of thought) that one is not concerned with knowing or thinking about the nature of quantum objects and processes, the existence of which it is compelled to infer from quantum phenomena. (The latter, we recall, are defined by Bohr as something that is registered in the measuring instruments.) Instead, it implies that, thus interpreted, the theory "*in principle* exclude[s]" the possibility of knowing, speaking, or thinking about their nature.[64] All that one can, in this view, conclude about quantum objects and processes is that they or, more accurately (I shall comment on this qualification presently), something in nature that compels us to introduce such objects exists and is responsible for certain phenomena that quantum mechanics or this interpretation can conceptualize, describe, explain, or use.

It would, accordingly, be difficult to conclude, as some authors have done recently in the context of quantum information theory, that quantum mechanics is only about information and manipulation of information, and that it is not concerned with quantum objects and their behavior, and hence is not quite physics. In the present view, quantum mechanics is a form of information processing (from one measuring device to another) rather than physics *only* if one thinks of the latter, as Einstein or Schrödinger did, on the model of classical physics or even relativity, or, as will be seen below, more broadly, if one thinks of knowledge and thought in (epistemologically) classical terms.[65] By contrast, the argument offered here is that, while quantum mechanics enables this type of information processing and thus enables quantum information theory, it is also,

---

[64] This claim is related to but is, generally, stronger than that of the impossibility of a joint assignment of both conjugate measurable qualities (as preexisting measurement) to a given quantum object at stake in the EPR-type arguments and Bell's and related theorems, as considered earlier.

[65] For this type of argument, one might refer to the works by A. Peres (Peres 1993), C. A. Fuchs and co-authors (e.g., Fuchs 2001, Fuchs 2003, Fuchs and Peres 2000) and to several others, including the present author (Plotnitsky 2002).

fundamentally, a form of physics insofar as it is about quantum objects and processes. It does, however, see the latter as something about which, as such, no information is possibly obtainable, even though and, it appears, because these objects and processes are productive of a special kind of information. Or more accurately (since this information itself qua information is classical), they are productive of information that could be generated and transmitted in a very special (such as the EPR-like) way. That is, this information is generated and transmitted by means of quantum physics but not by means of classical physics, although the latter can and, within certain limits, must be used in working with this information, specifically by transmitting this information (once it is generated) through classical channels. In other words, this information can be transmitted by both means, classical and quantum, but could only be created by quantum means. Accordingly, quantum mechanics is not reducible to quantum information theory, however much the latter may contribute, and it does, to our understanding of the physical, mathematical, and philosophical foundations of quantum mechanics.

As discussed earlier in this study, a potential difference between quantum objects, as the objects of quantum mechanics or of its interpretation, and the actual constituents of nature to which they may be related is important and must be kept in mind throughout this chapter as well. In particular, this view of the situation establishes the following important epistemological and philosophical point. Quantum mechanics (interpreted) as complementarity is, on the one hand, an idealization of nature, which, in principle, allows for otherwise idealized interpretations of quantum mechanics itself or for alternative theories of the phenomena in question. (In the present view, again, nothing other than such interpretations is possible.) On the other hand, it also implies that whatever is responsible in nature for this phenomena or this idealization may be unknowable even as unknowable or inconceivable even as inconceivable. For most practical purposes, especially as far as physics is concerned, one can provisionally suspend the question of nature itself and view a given field of nonclassical objects and their behavior (nonclassical processes) as something defined by the corresponding nonclassical theory itself, rather than as objects of nature. Thus, quantum objects and processes are provisionally seen as defined by quantum mechanics in a given nonclassical interpretation, say, Bohr's complementarity in its nonclassical version, rather than as certain objects found in nature. Indeed, the predictive aspects of quantum mechanics are unaffected by this bracketing of nature, since, in this view, the predictions in question concern only certain effects manifest in measuring instruments and described by classical physics and classical epistemology. In order, however, to rigorously establish the epistemological status of complementarity or any interpretation of quantum phenomena and quantum mechanics, one must consider its relation to a given frame or frames of reference. These frames of references are defined by nature or by the relationships between the theory and nature, along the lines discussed earlier in this study. That is, in the case of complementarity, one deals, on the one hand, with certain experimentally observed features of nature (e.g., certain effects manifest in measuring instruments) and, on the other, with certain unobserved and possibly (and possibly in principle) unobservable features of nature reflected in a theorization of quantum objects and processes as indescribable and even inconceivable. This is how the ultimate constitution of nature would be idealized by quantum theory in this interpretation. A certain process of establishing a proper frame of reference and a proper theoretical idealization of the objects considered is, as we have seen, also found in classical theories, such as classical physics. In this case, however, one obtains a different relation between nature and our theoretical idealizations, given the *descriptive* character of these idealizations. By contrast, nonclassical objects or their mode of existence may not be conceived of in any specific form available to our thinking, beginning with those attributes of (wave or particle) motion that define classical physics, but ultimately extending to all conceivable attributes. Accordingly, the term "existence" or any concept it would designate, or any other term or concept, named or not, that would refer to quantum objects or their behavior, "quantum" and "object" included, is ultimately inapplicable. Such terms or concepts may retain their provisional value in helping us to think, work, and communicate.

This view also appears to manifest itself in the statement to the effect that "there is no quantum world," famously attributed to Bohr (Peterson 1985, p. 305). I shall not discuss this much-debated statement here, in part because of the usual difficulties one encounters in interpreting or relying on such reported statements. It may be noted, however, that, in accordance with the nonclassical view just outlined, this statement need not be seen as implying that certain objects responsible for our, perhaps necessary, use of quantum mechanics do not exist. Quite the contrary; the independent existence of such entities is crucial for Bohr's argument concerning complementarity, in particular, as discussed earlier, in order to maintain the locality of quantum mechanics in view of the EPR experiment. Instead, the point is that this existence itself cannot be described or conceived by means of such ideas as "quantum" or "world," or even "existence" and "is" (however these are defined).

The preceding discussion makes clear that nonclassical epistemology can be seen as extending from but also moving us beyond the limits of the epistemology developed by Kant, in particular, as defined by his famous conception of things in themselves, as Bohr indeed noted in "Discussion with Einstein." As he said:

> Both in relativity and quantum theory we are concerned with new aspects of scientific analysis and synthesis and, in this connection, it is interesting to note that, even in the great epoch of critical philosophy in the former [the nineteenth] century, there was only a question to what extent *a priori* arguments could be given for the adequacy of space-time coordination and causal connection of experience, but never question of rational generalizations [such as that of

> complementarity] or inherent limitations of such categories of human thinking. (*PWNB* 2, p. 65)

Bohr's invocation of "space-time coordination and causal connection of experience" echoes his original conception of complementarity as the mutual exclusivity of these particular features of physical description in the Como lecture. His point itself, however, is aimed at indicating that nonclassical epistemology moves beyond that of Kant and critical philosophy, in particular insofar as, prior to quantum mechanics, the ultimately causal nature of physical processes and, by the same token, the possibility of establishing or assuming, at least in principle, causal connections between space-time events was not in doubt. Nietzsche might have been an exception, perhaps a unique exception, at least in philosophy (one can think of certain literary figures who entertain this type of questioning already in the wake of Kant's philosophy). Nietzsche appears to have sensed this radical potential, and he took it to the nonclassical limit in his own philosophy. On his way to proclaiming, "Long live physics!" in the passage cited above, Nietzsche argues that Kant ultimately failed to explore the epistemological potential of things in themselves and retreated to a more classical view of the world (now human world), more classical than the things in themselves would rigorously allow one to have (Nietzsche 1974, pp. 264-265). Nietzsche does, however, give Kant major (and justly deserved) credit in the same book: "*Kant's* tremendous question mark that he placed after the concept of 'causality'—without, like Hume, doubting its legitimacy altogether. Rather, Kant began cautiously to delimit the realm within which this concept makes sense (and to this day we are not done with this fixing of limits)" (Nietzsche 1974, p. 305). Unlike Nietzsche himself, however, Kant ultimately failed to become a "physicist" of the human world, which, Nietzsche argued, requires a different, ultimately (in the present definition) nonclassical understanding of the human world. The physical world needed quantum theory to compel us to move in this direction just a few decades later and to rethink the limits of causality in question in Nietzsche's comment.

In particular, while unknowable, things in themselves are still thinkable, including by means of such categories, and, thus, would, in the present view, be theorized as *classical* (in the sense to be defined presently). In Kant's words, "even if we cannot *know* these same objects as things in themselves, we at least must be able to *think* [of] them as things in themselves" (Kant 1997, p. 115). It is worth citing Kant's passage from his first critique, *The Critique of Pure Reason*, where this statement occurs, more extensively:

> We have no concept of the understanding and hence no elements for the cognition of things except insofar as an intuition can be given corresponding to these concepts, consequently [...] we have cognition of no object as a *thing in itself*, but only insofar as it is an object of sensible intuition, i.e., as an

> appearance [phenomenon]; from which follows the limitation of all even possible speculative cognition of reason [*Vernunft*] to mere objects of *experience*. Yet the reservation must also be noted, that even if we cannot *cognize* [*kennen*] these same objects as things in themselves, we at least must be able to *think* [*denken*] [of] them as things in themselves. To *cognize* an object, it is required that I be able to prove its possibility (whether by the testimony of experience from its actuality or *a priori* through reason). But I can *think* whatever I like, as long as I do not contradict myself, i.e., as long as my concept is a possible thought, even if I cannot give any assurance whether or not there is a corresponding object somewhere within the sum total of all possibilities. But in order to ascribe objective validity to such a concept (real possibility, for the first sort of possibility was merely logical) something more is required. This "more," however, need not be sought in theoretical sources of cognition; it may also lie in practical ones. (Kant 1997, p. 115; translation modified)

For example, when we think of our bodies as having a certain shape or organization, defined by such organs as the head, the arms and the legs, and so forth, we think of it on the basis of appearances. The very concept of the body is defined by this way of looking at it, possibly with inner organs, such as the heart, the liver, the brain, and so forth, added on. When, however, we think of the body as constituted by atoms or elementary particles, even if we think of the latter more classically, we think of the body as a (material) thing in itself. (The constitution of the body at other levels, physiological, biological, or chemical, may be seen as more knowable.) It is worth noting that the nonclassical view itself of quantum objects does not change this status of the body as thing in itself. For, while it is not knowable at this level, it is still *thinkable as* ultimately constituted nonclassically (along with everything else in nature), even though *the actual character of this constitution itself* is, by definition, not thinkable. That is, such is the case unless we consider the body or (this has been proposed by H. Umezawa and R. Penrose, among others, as in Penrose 1994) at least the brain as a quantum system and thus make it nonclassically unthinkable at the ultimate level. This view, however, remains conditioned by a particular, nonclassical interpretation of quantum objects, such the one offered here. For one can (as does Penrose) take an epistemologically classical view of quantum objects and processes.

Kant's qualification concerning practical sources of cognition could apply to most actual practices of quantum theory, when we, for practical reasons, apply classical physical features and attributes (particles, waves, positions, motions, etc.) to quantum objects. On the other hand, quantum phenomena appear to make it very difficult to find *theoretical* sources for applying such or any conceptions to quantum objects and processes, and in a nonclassical interpretation, such as Bohr's complementarity (at least in its post-EPR version), such an application is strictly impossible, is "*in principle* excluded." In such an interpretation, whatever they may relate to in nature, quantum objects and processes are rigorously placed beyond the limit of all conception and thinking, and thus beyond the limit of Kant's things in themselves or objects thus designated. (As noted above, it follows that the terms "quantum," "objects," or "processes," or any other terms are no longer rigorously applicable either.) This *placement* itself is of course justified theoretically, just as is Kant's introduction of things in themselves.

Kant's argument (important to all of his critical philosophy, in particular, along with the first *Critique*, to the second *Critique*, *The Critique of Practical Reason*) throws additional light on Einstein's discontent with quantum mechanics, and in part explains why Bohr thought that Einstein adopted Kant's view in his assessment of quantum mechanics (Faye 1991, p. ix). This link to Kant may be unexpected (especially if one forgets that reading Kant made a deep impression on young Einstein and, I would argue, affected all of his philosophy and some of his physics) but is revealing. I am not saying that Einstein's overall epistemology is identical to that of Kant, and Einstein was right to note, as he did on several occasions, the differences or at least complexities of the relationships between their respective epistemological and philosophical views (e.g., Schilpp 1949, p. 674). Indeed, as will be seen presently, Einstein's philosophy may be seen as closer to Hegel than to Kant, which proximity also further (and arguably better) explains his negative attitude toward quantum mechanics. My point is instead to stress the significance, apparent in this particular point, of Kant's ideas for Bohr and Einstein alike, and for the confrontation between them. The particular Kantian aspect of Einstein's view of quantum mechanics and complementarity in question is as follows.

We recall that Einstein admitted that "to believe" that quantum theory offers an exhaustive description of individual quantum phenomena, while suspending any possible description of quantum objects and processes themselves, "is logically possible *without contradiction*." However, he saw this view "as so very contrary to [his] scientific instincts that [he could not] forgo the search for a more *complete* conception" (Einstein 1936, cited in *PWNB* 2, p. 61, emphasis added). Given Kant's passage just cited, one can see this view as admitting the success of quantum mechanics as concerns its capacity to provide a logically consistent *practical* justification for its use as a set of effective predictive algorithms it provides, while arguing for our failure to offer a theoretical justification for the theory. In particular, it appears impossible to find a theoretical justification of the type classical physics and then relativity (both special and general theory) were able to offer us as descriptive idealizations of the physical processes they consider. The ultimate constitution of nature itself may still be seen in terms of Kantian things in themselves, but the difference between a practical justification, specifically for the purposes of correct predictions, in the case of quantum mechanics, and a theoretical justification, in terms of descriptive idealization, in the case of classical physics or relativity, is decisive. There are, again, certain complications in the case of relativity as well. Einstein was aware of these complications but (perhaps too optimistically) did not appear to view them as insurmountable as those of quantum mechanics and, accordingly, as requiring a replacement of relativity with an alternative theory. According to Bohr, in the case of quantum mechanics, such a theoretical justification, in terms of our understanding of the nature and behavior of quantum objects themselves, is not only currently unavailable but may well be, and in his (accordingly, nonclassical) interpretation is, "*in principle* excluded."

It should be noted that it is possible to see the justification in terms of idealized descriptive models as merely a practical one even in classical physics or relativity. In other words, similarly to the case of quantum mechanics in Bohr's view, it is possible to only see such models as *predictive algorithms* rather than idealized descriptions of actual physical objects and processes. It is difficult to attribute this type of view to Einstein, who appeared to think, as did Schrödinger himself, along the lines of what Schrödinger calls the classical ideal in physics, whereby our models and, hence, our thought maintain and progressively improve their connections with reality by way of idealized descriptions such models provide (Schrödinger 1935, *QTM*, pp. 153-154). This type of nonrealist view is, however, found in literature, especially in the case of relativity (Butterfield and Isham 2001).

Even if one adopts this view, however, there would still be a fundamental difference between relativity and quantum mechanics as complementarity, insofar as the predictive algorithms enabled by the formalism of quantum mechanics cannot be seen in terms of idealized (geometrical) models of the classical type. To make classical physics or relativity epistemologically nonclassical, it would not be enough to merely renounce the view that the causal models they use are descriptive idealizations of reality and only to see them as offering predictive algorithms for the outcomes of relevant experiments. It would be necessary to offer interpretations of both in which it would be rigorously impossible to see these models as descriptive idealizations. I am unfamiliar with interpretations of either classical physics or relativity that are fully nonclassical in this sense. Indeed, while the situation is, again, more complex in relativity, in which case some at least question the possibility of seeing it in terms of idealized descriptive models of physical reality, classical physics usually is, including by Bohr, interpreted along epistemologically classical lines of idealized descriptive models.

As I have argued in this book, quantum mechanics as complementarity disallows such causal models, even if one only wants to see them as predictive algorithms rather than descriptive idealizations. For, in this view, the formalism of quantum mechanics does not conform to or allow for an interpretation in terms of such models. In other words, the mathematical formalism of quantum mechanics does not and, it appears, cannot yield phenomenalizable physical models. Or, more accurately, it does not offer mathematical models (causally) representing objects and motion, even if the latter are seen as ideal or, one might say, "fictional," rather than real or approximating reality. Accordingly, in this case it is not a question of a possible alternative interpretation of our models either as descriptive idealizations or predictive algorithms. Both the mathematical formalism and quantum objects and processes (as the objects of quantum mechanics) are rigorously placed beyond any phenomenal conception, which could allow for an ideal model of the motion of physical bodies. All we have is a mathematical (algebraic) conception of the formalism that enables our predictions concerning the outcome of certain specific experiments or phenomena. We have and can have no conception concerning quantum objects and processes as considered by the theory, which may or may not correspond to the way they are in nature, that is, in our ultimate capacity to relate to nature. Conceptual, linguistic, mathematical, experimental or still other forms of objectivity are retained (it would be difficult to do physics in its disciplinary sense otherwise) in Bohr's argumentation concerning complementarity and, as noted earlier, are maintained by Bohr throughout. They are, however, transferred to the level of measuring instruments or other macro-objects, described, whenever their physical description is necessary, by means of classical physics.

With the preceding discussion in mind, one may define as *classical* an account that would, at least in principle, determine all of its objects (which one may also call "classical") as knowable or, analogously to Kant's things in themselves, at least as thinkable or subject to one or another type of modeling. As just noted, classical physics may be and usually is seen or interpreted, as it was by Bohr, as epistemologically classical in this sense, and as idealizing the physical objects in nature, such as planets moving around the sun, accordingly. The epistemologically classical character of classical physics is essential to Bohr's interpretation of quantum phenomena and quantum mechanics as complementarity, given both the irreducible role of measuring instruments there and the description and describability of these instruments by means of classical physics. In this interpretation, such a classical physical description, again, applies *only to certain parts* of these instruments. Certain other parts or strata of these instruments, specifically those by means of which they interact with quantum objects, must be seen as quantum, and hence in turn as subject to nonclassical epistemology. As Bohr notes, the ultimate constitution of our measuring instruments, including those parts that are described classically, may of course be quantum, as may be the ultimate constitution and, hence, the ultimate description or, as the case may be, undescribability of all classical objects. As he also notes, however, "this circumstance is not relevant for the problems of the adequacy and completeness of the quantum-mechanical description in its aspects here discussed," that is, primarily, the nonclassical aspects of Bohr's interpretation of quantum mechanics as complementarity (*PWNB* 2, p. 51). As we have seen in Chapter 5, this circumstance appears to become relevant in quantum field theory. In any event the objects of nonclassical theories, such as quantum objects in quantum mechanics in a nonclassical interpretation, are *irreducibly* unthinkable, even as objects or things in themselves.

It follows that nonclassical theories suspend realism at the ultimate level that they consider, if we understand by realism the possibility, at least in principle and by way of idealization, of mapping or conceiving of the properties and behavior of a given entity or system, whether the latter describes or approximates nature or not. (Thus, one might try to conceive of quantum objects and behavior on the model of classical physics.) The suspension of causality, at the same level, is, again, an automatic consequence: if one cannot conceive of nonclassical objects and their behavior at all, one cannot claim this behavior to be causal. Quantum theory and other nonclassical theories, however, suspend causality at other levels as well. Classical theories are, by definition, realist in the sense just explained, which allows for the possibility of knowledge or at least conception concerning all of their objects, in physics, usually seen as an idealized model of natural objects and their behavior. Indeed, in this view, classical and realist theories are the same. Many of such theories are also causal, although, again, not all of them are deterministic, that is, allow us to make sufficiently exact predictions concerning the outcome of the experiments we set up of the type we can do in classical mechanics. Neither classical statistical theory nor chaos theory is deterministic, although, as explained earlier in this study, for different reasons. Both, however, are causal. Indeed, we do not appear to have noncausal theories that are subject to a realist description and may not have a causal character, if Ludwig Wittgenstein is right in thinking that we cannot conceive of processes that are not causal (Wittgenstein 1985, p. 179). This is a view that Bohr appears to share, as is suggested by many of his arguments and even titles, in which the term causality occurs often. Wittgenstein makes his observation in 1924, around the time of the emergence of quantum mechanics, when the problems of causality in quantum physics were already apparent from the old quantum theory.

It also follows from the preceding argument, however, and indeed from the very definition of nonclassicality that nonclassical theories contain classical and even strictly knowable strata. They must do so, given that the existence of unthinkable objects is rigorously derived by a given nonclassical theory, as opposed to being merely postulated. For such a rigorous derivation is not possible otherwise than on the basis of something that could be and is known, even though it must also be seen as impacted by what is not and cannot be known or even thought of. We *know* of the existence of nonclassical objects or, again, something in nature that compels us to introduce them, and define and, hence, *know* (rather than only think) them to be unthinkable through their effects upon objects or phenomena that we know, and only through these effects. Accordingly, as explained earlier in the context of quantum mechanics and complementarity, nonclassical knowledge only concerns effects produced by certain nonclassical objects in question in a given nonclassical theory upon other, knowable and hence classical, objects. At the level of such effects, realism and classical thinking, including objectivity, would apply.

Causality appears difficult to claim even at the level of effects in both quantum theory and in the nonclassical view of the mind by virtue of nonclassical and, hence, by definition, acausal processes intervening between and affecting classical events in question.

Thus, the existence of something in nature, from which such objects are inferred, is manifest in certain particular effects of the *interaction* between that something with that part of the physical world that we can observe or know (even if only in some idealized way), the "classical" world. These may be, for example, those effects that are observed in the double-slit experiment or particular types of correlations between certain events in quantum physics. Accordingly, as Bohr argues throughout his writings on quantum mechanics, this *interaction* itself is irreducible in our dealing ("interaction") with quantum objects; that is, they cannot be *theoretically taken into account* apart from this interaction by means of our measuring instruments or other macro-objects that perform an analogous function on their own accord. I speak of *theoretically taking such objects into account* because in this view it is, again, impossible to speak of *considering* (describing, explaining, or conceiving of) them in any rigorous sense. Bohr, as this study has (repeatedly) stressed, repeatedly speaks of "the *impossibility of any sharp separation between the behavior of atomic objects and the interaction with the measuring instruments which serve to define the conditions under which the [observable] phenomena [in question in quantum physics] appear*" (Bohr 1949, *PWNB* 2, pp. 39-40). This impossibility compels us to think of these "objects" as nonclassical, and, once again, Bohr's statement must still be seen as demarcating atomic or quantum objects theoretically rather than, at least at the level of the objects in question, describing nature itself at the quantum level, which may, accordingly, be ultimately unthinkable even as unthinkable in this view. But this view also allows for the possibility that the quantum constitution of nature might be theorized otherwise, via quantum mechanics or other theories. At the level of the effects in question, we can speak of an idealization of nature along the lines of classical physics. Nature, however, or something in nature compels us to a nonclassical way of thinking of the quantum.

3. EPISTEMOLOGY AND INVENTION OF CONCEPTS: BOHR AND EINSTEIN BETWEEN KANT AND HEGEL

I argue here that in his epistemological thinking, Bohr may be seen as a Kantian thinker, although his ideas also radicalize Kant's epistemology. He may be seen as a *more* Hegelian thinker as concerns his invention of new philosophical concepts and conceptual architectures, which enable us to confront this epistemology and indeed put it to work. The capacity of our thought to do so largely grounds Hegel's response to Kant as well, although Hegel's thought has its nonclassical dimensions or anticipations as well, just as the significance of the invention of concepts is not missed by Kant. As in the practice of all major philosophy, the practice itself is, certainly, found in Kant's work, with such concepts as things in themselves or phenomena as, arguably, the primary examples. These are indeed these concepts that were radicalized by Bohr into his concepts of quantum objects and phenomenon, respectively, which also makes him indebted to Kant's concepts. Hegel, however, centers his view or his concept of philosophy around the concept of concept (itself a new concept, conceived along the lines sketched above) and the invention of new concepts much more deliberately and pointedly than Kant or arguably anyone else before him. From *The Phenomenology of Spirit* (Hegel 1977) on, all meaningful thought processes and, with them, all meaningful historical processes, for example, those defining the history of science, are governed and defined by their conceptual organization (*die begriffene Organization*).

It is important that the architecture of concepts, philosophical or scientific, including mathematical, in Hegel is governed by their reciprocal constitution of the type considered here, although, for Hegel, philosophy defines the ultimate structure of all fundamental concepts. Thus, in his long (almost infinite) "Remark" on "The Specific Nature of the Notion of the Mathematical Infinite" in *The Science of Logic* (Hegel 1990, pp. 240-313), Hegel gives mathematics a special role in the history of thinking in terms of concepts, even in philosophy, rather than only physics. At stake in particular is the concept of infinity and, or as, continuity, which are, as we have seen, crucial to the philosophical conceptuality and the epistemological problematic of modern physics, and which significantly shape the epistemological difference between classical and quantum physics. According to Hegel, "the character of the mathematical infinite and the way it is used in higher analysis corresponds to the Concept of the genuine infinite" (Hegel 1990, p. 244; translation modified). Hegel juxtaposes this concept both to the concept of "the metaphysical infinite" or to Kant's concept of the infinite, to both of which the mathematical infinite of the differential calculus is, accordingly, superior (Hegel 1990, pp. 241-243). The mathematical infinite may not reach this concept of the genuine infinite, that is, the concept of the infinite that can be developed in philosophy, such as that of Hegel, which (Hegel claims no more for himself) approaches it, although, Hegel argues, not that of Kant. This argument makes Hegel's "Remark" one of the important junctures of his confrontation with Kant. (This critique of Kant is of further interest because Kant, too, gives mathematics a special role in philosophical thinking, although I can only mention this point without pursuing it here.) The significance of the mathematical infinity and, accordingly, of calculus, as its first mathematically rigorous elaboration or construction, remains paramount, conceptually and historically. This is all the more so because the concept of the infinite defines Hegel's overall view of philosophy and its history (these are, as I said, indissociable), conceived or, again, *conceptualized* in terms of the development of concepts. In his "Remark," Hegel specifically considers Euler's and Lagrange's concepts of differential calculus, and their implications for the difference between the geometrical and algebraic approach to calculus and, at least by implication, to physics.

Hegel's analysis there has considerable pertinence to the present discussion and this study as a whole, in particular as concerns the epistemology of the infinite or the relationships between physics and mathematics, or algebra and geometry in both, as discussed earlier. Indeed, Hegel's critique of Kant's view of the infinite helps us to realize that, in his language, the *genuine* (that is, truly rigorous) epistemology of the infinite may ultimately need to be nonclassical. Hegel's analysis cannot, however, be considered here, in part because, while suggestive and deep philosophically, it is not unproblematic, and carefully sorting it out would require too much space and will take us well beyond the limits of the present chapter or even this study. My main point here is the significance of conceptual invention and the role of mathematics in this process even in the philosophical sphere, let alone in physics, and it is this significance and the very concept of concept, as considered here, that define Hegel's philosophy. I shall, accordingly, restrict myself only to a few pertinent points related to this analysis or its implications.

The (classical) epistemology of classical physics, including as allowing for the possibility of geometrical representation or intuition (*Anschaulichkeit*) there, at least, again, at the level of idealized models, is essentially linked to the possibility of representing the processes in question there in terms of differential functions and differential calculus. In classical physics or relativity, mathematical and physical continuity work together and, one might argue, correlatively, allow for classical epistemology. By contrast, in quantum mechanics or, at least, complementarity the mathematical continuity serves physical and epistemological discontinuity and, or as, unrepresentability or inconceivability, which leads to nonclassical epistemology. Certain philosophical intimations of this problematic emerge in Hegel's analysis in his "Remark" and throughout this chapter of *The Science of Logic*, prophetically, even if coincidentally, entitled "Quantum." It is coincidental insofar as there was no quantum physics at the time, although Newton's corpuscular theory of light, well known to Hegel, could have played an, at least, analogous role. It may be seen as less coincidental historically, insofar as the question of the relationships between continuity and discontinuity in mathematics and physics, or philosophy is at stake. These relationships are given a very special shape in quantum mechanics, especially as complementarity, first, insofar as continuous mathematics enables correct numerical statistical predictions concerning certain discrete phenomena and, second, insofar as these phenomena themselves are given a complex physical constitution, rather than being physically discrete. The numbers defining the predictions in question do form discrete sequences of individually random but collectively correlated events. This is quite an extraordinary conceptual and physical architecture.

This is why I argue that, while Bohr's epistemology is closer to that of Kant, he is Hegelian as concerns the significance of concepts and especially of the invention of new concepts in his approach to quantum mechanics, although Kant's and Hegel's views on both subjects significantly overlap in turn, more than Hegel likes to think. As indicated earlier in this study, to some degree (but only to a degree), the initial, Como, version of complementarity may be seen as more philosophical, while the subsequent versions (it appears, all of them) may be seen as more empirical as concerns the key complementary features or phenomena concerned. The Como version is defined by the two main complementarities (in the narrow sense of the mutual exclusivity of certain entities or features of description, a concept retained throughout all of Bohr's work). These complementarities are more conceptual, philosophical in character and were replaced in Bohr's subsequent thinking by the complementarities of certain alternative experimental arrangements (suited for certain mutually exclusive types of measurements, such as those of position or momentum). The first is the complementarity of the space-time coordination and the claim of causality, which, Bohr adds, "[symbolize] the idealization of observation and definition respectively" (*PWNB* 1, p. 55). This qualification gives rise to the second complementarity, again, more conceptual or philosophical in nature, the complementarity of the two idealizations in question, that of observation and that of definition. On the other hand, the difference in question is only that of degree, since, while Bohr's post-EPR thinking aims at a more empirical grounding of complementarity, new philosophical concepts are introduced as well, most especially those of phenomenon and atomicity, as discussed above. Conversely, while the main complementarities found in the Como lecture, just mentioned, may be seen in more philosophical terms, they carry a crucial physical significance by virtue of the juxtaposition with classical physics that they establish, and, in the case of the second complementarity, by virtue of the relationships between experiment (observation) and theory (definition).

In contrast to Bohr, who couples his invention of concepts (or his critique of classical concepts) to nonclassical epistemology, Einstein, while he also took a Hegelian view, just sketched, concerning the significance of concepts, concepts in general and new concepts in particular, remained close to the epistemologically classical aspects of both Kant's and Hegel's thought. Both philosophies, again, have nonclassical ingredients as well, which were perhaps resisted more by Einstein than by Kant and Hegel. A good example is Einstein's famous argument that our concepts and theories decide what could be observed. Interestingly, Martin Heidegger makes a similar point assessing Galileo's project, in part against positivism (again, in accord, with Einstein's maxim in question). As he says: "Modern science is experimental because of its mathematical project. The experimenting urge to the facts is a necessary consequence of the preceding mathematical skipping (*Überspringen*) of all facts. But where this skipping ceases or becomes weak, mere facts as such are collected, and positivism arises" (Heidegger 1967,

p. 93). Einstein's argument, we recall, impressed Heisenberg and guided his work on his paper on the uncertainty relations, as against a more positivistic view of basing one's theory of the quantities that are in principle observable adopted in his original paper introducing quantum mechanics (but with qualifications given in Chapters 1 and 2). Einstein himself was previously, but (guided by the principles that led to general relativity) not by this point, an adherent of Mach's philosophy, largely responsible for the principle of founding a physical theory on the quantities that are in principle observable. Einstein's new principle appears to have its origin in his work on the general (rather than special) relativity theory. Einstein's insight is significant insofar as it leads to a questioning of the uncritical use of the idea of observation. He argues, in a Hegelian vein, against the empiricist or positivist "philosophical prejudice," which "consists in the belief that facts by themselves can and should yield scientific knowledge without free conceptual construction."[66] He adds, however: "such a misconception is possible because one does not easily become aware of the free choice of such concepts, which, through success and long usage, appear to be immediately connected with the empirical material" (Einstein 1979, p. 47). Accordingly, this view and, with it, the form that this principle of the theoretical definition of observation takes, or, at least, the way the principle was used (for example, in general relativity) by Einstein himself, entail a form of realism, albeit of a complex and indirect (rather than naïve) kind. In particular, in his view now

[66] Hegel noted that empiricism forgets at the very least that it uses the word "is." In other words, empiricists take this use for granted, as opposed to seeing the concept of "being" as part of our conceptual thinking that shapes all of our observations. The latter view may be seen as the ultimate extension of the type of view concerning the relationships between observation and theoretical definition in physics advocated by Einstein. Some of these complexities are reflected and, along with both Einstein's and Heisenberg's views on the subject, might be among the sources of Bohr's conception of the complementary nature of "observation" and "definition" in quantum mechanics, manifest in the complementarity of the space-time coordination and (the claim) of causality, in his Como lecture. As explained earlier, Bohr abandons these conceptual complementarities in his subsequent writings in favor of more experimentally defined complementary features, such as those that are manifest in alternative measuring arrangements. This does not mean, however, that Bohr shifts to a more positivist view or no longer takes into account the subtleties of the relationships between various forms of determinations of quantum phenomena in developing his concepts and his interpretations of quantum phenomena and quantum mechanics as complementarity. As I have stressed throughout, even though and because quantum objects are indescribable and even unthinkable, their existence, including as unknowable and unthinkable, is essential to complementarity as an interpretation of quantum mechanics. The mathematical-experimental practice of quantum theory may allow for either view (positivist or that of Bohr), but one's interpretation of quantum phenomena and quantum mechanics is essentially affected and even defined by the differences between these views.

access to reality is possible otherwise than through a mediation by concepts (yet another Hegelian process). According to this view, however, a physical theory should, ideally, also enable one to at least approximate the physical reality, specifically in terms of physical properties of the objects in question and processes occurring in space-time. As noted above, while it is in principle possible to view the models employed as predictive algorithms rather than descriptive idealizations, it is difficult to see Einstein's attitude in these terms. For him conceptual thinking on imagination in physics links us to reality and even approximates reality. In quantum mechanics as complementarity there is, again, no such alternative, since there are no idealized models of physical types, however ideal, attached to the formalism but only the mathematics of this formalism that offers a set of effective predictive algorithms for the outcome of the relevant experiments.

Einstein's view just described is, again, closer to Hegel than to Kant and in part explains why he sometimes dissociates his own views from those of Kant, although there is no strong supporting evidence for a specifically Hegelian genealogy of Einstein's view. Hegel's influence upon the whole of nineteenth-century thinking, including in physics or elsewhere in science (for example, in evolutionary theory) was pervasive, and these affinities with Einstein's thinking, mediated through other sources as they may be, are hardly surprising. Numerous traces of Hegelianism, even if not of Hegel's own thought, are found in Einstein's philosophical positions and arguments; and Hegel's analysis certainly develops the *concept* of mediation and mediated determination by way of concepts that Einstein wants in physics, an attitude that may, accordingly, be seen as Hegelian. From the *Phenomenology* on, Hegel undertakes an unprecedented exploration of the problematics of conceptual mediation of reality. Indeed, he may be shown to ask and in great measure to answer most of Einstein's philosophical questions and within limits that are broader than those envisioned by Einstein and that may indeed be required in order to give classical physics the kind of philosophical grounding Einstein wants. Once we move to quantum mechanics this grounding may no longer be possible, and, in the nonclassical view, it is rigorously impossible, even though and because the work and the very invention of concepts remains as crucial as ever.

However they may be determined conceptually or historically, Einstein's *epistemological* views, as just sketched, arguably best help to explain Einstein's negative attitude toward quantum mechanics, which might seem puzzling, especially given that Einstein's view, just discussed, applies to theoretical innovations in physics or elsewhere. Thus, the ideas of both Kant (along the lines explained in Section 2) and Hegel philosophically shape this attitude. Quantum mechanics does lead to new concepts, such as complementarity (at least in the narrow sense), at least if it is considered as a philosophical concept. As we have seen, however, the situation is more complex as concerns the status of these concepts as *physical* concepts, insofar as classical physical concepts, are, with Bohr and Heisenberg, again, seen here as representing a suitable and suitably mathematizable refinement of our everyday ideas concerning material bodies and their motion. As discussed earlier, the radical unthinkability of nonclassical objects does not mean that certain "concepts" specific to a given nonclassical theory, for example, certain specifically quantum (i.e. not found in classical physics) concepts, such as "spin," cannot be introduced in quantum mechanics—quite the contrary. The question is to what degree, if any, we can conceptualize something like "spin" at the quantum level in terms of classical (that is to say, any) concepts, as opposed to defining the field of measurable effects associated with them and developing a mathematical formalism for predicting such effects. Both of these we can do rigorously. If anything, "spin," the famously inconceivable "angular momentum" (a useful metaphor borrowed from classical physics but ultimately inadequate to describe "spin"), is a good paradigmatic case of this situation. Rigorously, it is just a set of certain numbers obtained in measurement, which fact would explain why at some point (before quantum mechanics or even spin itself) Pauli said that numbers appear more real than electrons' orbits in atoms. Pauli's famous exclusion principle is best understood along these lines of thinking in terms of certain available or forbidden sets of numbers rather than a physical conception of the state of an electron in an atom. In this sense, no other than classical and in this sense "old" concepts are possible, or at least no concepts extending beyond a certain enclosure of classical conceptuality, within which the invention of new concepts continues to take place, similarly to the way it happened previously. The invention of the classical concept of motion is itself an earlier example of such an invention, which indeed took a while, with its arguably most significant stage taking place with Galileo and Newton, whose work made their mathematical form or their relation to mathematics possible.

With this view in mind, while (similarly, say, to Einstein's general relativity) based on mathematical innovations or new applications of mathematical theories in physics, physically quantum mechanics appears to rely fundamentally, if not exclusively, on a new use of old physical concepts, especially those of or related to motion. A different use and delimitation of such concepts, as against those in classical physics, and, it follows, their critique are necessary in quantum mechanics, for example, a limited (by the uncertainty relations) and necessarily complementary use of some of them. In Bohr's post-EPR scheme, moreover, these concepts are now only applicable to the measuring instruments involved, although this use, self-evidently, itself involves new concepts, physical and philosophical, such as that of complementarity (in the narrow sense). In earlier thinking of Bohr and of Heisenberg, these concepts could still be applied in a limited or, again, critically reassessed and re-delimited, or, as we sometimes say now, deconstructed, way to quantum objects themselves. At the same time and (in view of the potential inapplicability of our everyday concepts, however refined, at the quantum level) as a consequence, it does not appear possible to develop the concepts, the new concepts just mentioned included, through which we can describe or even conceive of quantum objects themselves and their behavior. The latter are thus unreachable by the theory, which fact breaks the possibility of any conceptual connection with or possible conceptual approximation of physical "reality" at the level of the ultimate objects considered by the theory. As discussed earlier, at most one is left with a practical, specifically predictive, justification of the theory, while not only (along more positivistic lines) abandoning all attempts to explain the ultimate constitution of nature, but arguing, nonclassically, that such an explanation or even any conception of this constitution is, or at least might be, beyond our reach, "*in principle* excluded."

This is of course not the kind of theory building that Einstein or his predecessors, such as Maxwell (as concerns Einstein's work in both electrodynamics and in the kinetic theory of gases), so effectively used previously and that led Einstein to his belief in the role of field theory in fundamental physics. This type of approach defines, to return to Schrödinger's language, the classical ideal in physics, with which quantum mechanics breaks, thus implying that this ideal may not do "justice to nature" (Schrödinger 1935, *QTM*, pp. 153-154). Nor is this the kind of theory building that Einstein or Schrödinger could have been expected to welcome. If anything, one would expect them to have had deep doubts and concerns about it or even to have deplored it, as both indeed did. In the case of quantum mechanics, especially if understood along the nonclassical lines of Bohr's complementarity, new concepts, by definition, no longer appear to bring us any closer to reality, or, as Einstein liked to say, to the "secret of the Old One," of the God who does not play dice. Or, again, at least, God, Einstein thought, would be extremely unlikely to set the universe up according to the ways in which chance works in quantum mechanics. If anything, these concepts place any "reality" at an unreachable, "infinite" distance, which no concept of reality can help us to travel and at which, therefore, the term itself, reality, loses any validity. It, again, remains valid and even indispensable within certain limits, the limits that I, accordingly, define as classical here and that are superceded by nonclassical thinking.

The concept of complementarity or the complex of concepts forming Bohr's framework is Bohr's greatest physical and philosophical invention. The centering of this framework around any given concept, say, complementarity as a mutual exclusivity of certain entities, is relative, since one can, at least partially, re-center it around other concepts, such as that of phenomena or that of atomicity, or one can give all these concepts equal centrality. These concepts, however, are correlative to and embody the radical epistemological features defining quantum phenomena and quantum mechanics in Bohr's interpretation. In other words, this interpretation is jointly defined by this epistemology and these concepts. By virtue of this co-definition, Bohr's thought on complementarity brings the Kantian and the Hegelian ways of thinking together by, on the one hand, retaining the epistemology that is analogous to and reaches beyond that of Kant's things in themselves, and, on the other, by introducing a set of concepts that relates to and, again, embodies this epistemology. It is true that the thought of Kant and, especially, Hegel and other post-Kantians (such as J. G. Fichte and F. W. J. Schelling) also responds to the epistemological difficulties, defined by Kant, with which nature and mind confront us. They do so by creating profound concepts that overcome these difficulties and bring us closer to an understanding of the ultimate workings of mind and nature, even not or even never quite reaching such an understanding. The main difference between their thought and that of Bohr, which defines Bohr's thinking as nonclassical, is that Bohr's concepts no longer aim at reaching the ultimate constitution, the ultimate *nature* of nature, but enable quantum mechanics to work under these radical conditions.[67] The situation is, I argue, generalizable to nonclassical theories elsewhere, and in the process, I also argue, the very nature of thought and knowledge is redefined, re*thought* so as to make the unthinkable and the unknowable essential, irreducible parts of thought and knowledge, as ultimately responsible for what we can think about and know.

Nonclassical knowledge and thinking are no less rich or deep than those of classical theories, which are part of nonclassical theories in any event. It is not a matter of epistemological preference (or prejudice) but of theoretical necessity that may compel us to classical approaches in some cases and nonclassical in others. Nonclassical theories do expand our understanding of the nature of fundamental explanation in science, philosophy, and other fields. The nonclassical attitude is not defeatist, even though and because it establishes certain uncircumventable limits upon how far our theories allow us to reach. For one thing, these are our theories and, hence, our thought that rigorously establish these limits. Secondly, and more importantly, nonclassical theories are productive of new thinking and knowledge, physical, mathematical, and philosophical, under these epistemological conditions and limits, as, following Bohr, I have stressed throughout this study. Quantum field theory is a primary example of moving beyond quantum mechanics but retaining nonclassicality. Indeed, when one says that nonclassical theories place their ultimate objects beyond any knowledge or even conception available to us, the terms knowledge and conception are used as classically conceived. One may, however, expand the conception of knowledge and thinking so as to allow for the inclusion of the nonclassical unknowable and unthinkable and to allow for knowledge and thinking to be conceived in terms of effects of this unknowable or unthinkable upon what is knowable and thinkable. This conception itself is of course still classical as a *conception*, and there is, by definition, no other way for us to conceive of anything rather

---

[67] I leave for the moment aside the question of whether the post-Kantians and their concepts handle these Kantian difficulties better than Kant himself did, which is the subject of complex interpretive arguments and (just about interminable) debates. My point here is not affected by these considerations, unless one argues that the thought of some among these thinkers reaches nonclassical limits. I am unaware of such arguments. Even if such a case were made, however, say, for Hegel, it would merely mean that an argument analogous to my overall argument for Bohr's nonclassical epistemology would apply to Hegel's thought, similarly to the way it would and, in my view, should apply to the thought of Nietzsche.

than classically. What is different is the *character* of knowledge and thinking found in nonclassical theories.

This type of argumentation is found throughout Bohr's work, in particular and especially significantly in his reply to EPR. There the nonclassical epistemology of complementarity is seen as a condition of the continuation of physics and hence of classical physics (hence, again, moving beyond it without leaving it behind), as a discipline, that is, in Galileo's language, as a mathematical science of nature. It also follows from Bohr's analysis there, that, if the epistemology of complementarity itself entails a new view or indeed a new philosophy of knowledge (epistemology) and possibly of nature itself (as at the ultimate level of its constitution inaccessible to human knowledge or conception), this argument directs us to a new philosophy of science. I shall discuss this last point in Section 5. Before I do so, however, I would like to reexamine from the (more) philosophical perspective here developed, Heisenberg's discovery of quantum mechanics, as discussed in Chapters 1 and 2, or to recast this argument concerning this discovery in a more general philosophical form by bringing together epistemology and the invention of concepts. Heisenberg's discovery is arguably the single most important event responsible for the emergence of this perspective. Or it would be if there could be such a single most important event, which is difficult and, it appears, ultimately impossible to maintain, even leaving aside the fact that this or all such events are never single events but are themselves historical developments. I shall also discuss in this section Heisenberg's critique (in Kant's sense) of the concepts of classical physics by means of quantum mechanics.

4. THE DISCOVERY OF QUANTUM MECHANICS AND THE CRITIQUE OF CONCEPTS IN HEISENBERG

Although Bohr, beginning with his 1913 theory of the hydrogen atom, arguably deserves the greatest credit for the development of nonclassical epistemology, at least in physics, part of my historical argument in this book is that the road to the nonclassical epistemology of quantum mechanics may be seen as initiated by Heisenberg's approach in his discovery of quantum mechanics. To return to Bohr's initial characterization of Heisenberg's theory: "In contrast to ordinary mechanics, the new mechanics does not deal with a space-time description of the motion of atomic particles. It operates with manifolds of quantities which replace the harmonic oscillating components of the motion and symbolize the possibilities of transitions between stationary states in conformity with the correspondence principle. These quantities satisfy certain relations which take the place of the mechanical equations of motion and the quantization rules" (*PWNB* 1, p. 48). As discussed in Chapters 1 and 2, and earlier in this chapter, to abandon, as Heisenberg did, "a space-time description of the motion of atomic particles," is not quite as radical a step as making an analysis of such motion or, yet more radically, any conception of quantum objects and their behavior "*in principle* excluded." This step was first taken by Bohr, and it took a while and some "help" from Einstein, whose arguments, especially those of the EPR type, compelled Bohr to make this step. Nevertheless, Heisenberg's step was decisive, short of the locality considerations involved in the EPR-type arguments and Bell's and related theorems (far away at the time), physically it takes us nearly as far, certainly beyond anything hitherto encountered in physics.

There is much philosophy of nature and of physics involved in this step as well. Most especially, the concept motion that previously defined mechanics is now replaced with (Bayesian-like) probabilistic estimations of individual events, with a concomitant introduction of a new (matrix) type of mathematical variables in an infinite-dimensional space, one of Heisenberg's great conceptual inventions in physics. We recall that the corresponding mathematical objects have been introduced earlier, but were reinvented, from physics, by Heisenberg. His invention of this "new calculus" for physics brings Heisenberg's achievements to the level of those of Galileo and Newton, or Einstein's tensor variables of general relativity.

As discussed in Chapter 1, the introduction of new mechanics in Heisenberg's great first paper on the subject may be seen as *arising* from, or at least as linked to, an extraordinary form of "vision" of the material constitution and, with respect to the viewpoint of classical physics, de-constitution of the data in question. I am, again, speaking here of the vision relating to the entities constituting the quantum data, such as those marks or traces that are found in spectra or in the dot-like patterns on the photographic plates in the double-slit experiment, rather than of the theoretical conceptualization of the quantum-mechanical situation, which is, however, ultimately made possible by this vision. The vision itself may be seen as physical insofar as it deals with a certain conceptualization of physical data, including, again, via relating these data to his matrix elements. This vision may, however, no longer be rigorously seen as a (mathematical) formalization as regards quantum objects and processes, in the way classical mechanics is as regards the objects and processes it considers. The mathematics of quantum mechanics is of course just as formal as that of classical physics and, given its Hilbert-space formalism, may even be seen as more abstract and complex than that used in classical physics, although the latter could be given a more mathematically abstract form as well (symplectic geometry, fiber bundles, and so forth). For the moment, however, I am speaking of *envisioning* physical processes and mapping them, which the mathematics of classical mechanics does and that of quantum mechanics, in this "vision," does not and indicates that it may not be possible to so. In contrast to the ultimate constituents of matter or the ultimate efficacious dynamics responsible for the emergence of the data in question, the (material) elements, marks, constituting these data are available to phenomenological apprehension. One does not, however, treat these marks, even their ordered collectivities, in terms of physics, as opposed to our phenomenological perception of them, say, as wave-like interference patterns of dots in the double-slit experiment. Instead, one, as it were, divests these marks or rather the processes of their emergence (an analysis of these processes is abandoned altogether, or even eventually seen as prohibited) from their classical and hence configurable conception of process, motion, etc. Indeed, one divests these marks and their emergence of anything that could possibly be mapped by a classical model, even though they do form configurations, or what can be so seen in certain circumstances—such as, again, the wave-like interferences pattern in the double-slit experiment, or a trace of a particle in a cloud chamber. The processes leading to the appearance of these marks must be divested of the possibility of being explained in classical terms and hence of their (apparently) manifest classical and, again, specifically geometrical configurativity. For example, these marks should not be seen either as points resulting from *classically conceived* collisions between "particles" and the screen or as forming a classically conceived wave pattern. Neither "picture" corresponds to what in fact occurs. At this stage, even the trace-like character of these marks and that they are traces of certain physical processes that are ultimately inaccessible to us are suspended, although this type of character will have to be given to these marks in order to explain them in quantum-theoretical terms. In sum, this vision is that of a suspension or removal of classical physics from considering the emergence of the data and phenomena in question.

This suspension is necessary and the vision that results is possible for the following reasons, apparent from the preceding argument of this study. The mathematical formalism of quantum mechanics, at least in Heisenberg's original view and more radically in Bohr's interpretation, does not describe or otherwise formalize the processes leading to such configurations of marks or traces, individual or collective, or for that matter any material physical process in the way classical physics would. It only enables correct statistical predictions concerning individual marks or configurations of such marks in certain circumstances.[68] Accordingly, in order for a theoretical formalization and interpretation of quantum physics to take place, the physical emergences of these marks have to be divested of any form of mathematical and specifically geometrical representation. Classical physics is largely defined by the possibility of such phenomenal representation, even if idealized, of the physical processes it considers, thus making them available to and, at least in part, reciprocally deriving them from human intuition (*Anschaulichkeit*). In quantum mechanics, in a nonclassical interpretation, this is no longer possible. This impossibility is reflected in the nonclassical nature of the interpretation of the traces constituting the phenomena or data considered and in the corresponding or correlative interpretation of the mathematical formalism of quantum

---

[68] We recall that the nature of quantum probability is in turn nonclassical, and is not defined, as in classical statistical physics, by, in practice, insufficient information concerning the systems that, in principle, behave classically.

mechanics as a predictive machinery for the numerical aspects of these phenomena. The efficacious processes themselves giving rise to these phenomena are beyond the reach of any pictorial visualization, intuition, phenomenalization, representation, conception and so forth. Accordingly, or even "beforehand," the origin and genesis of the manifest effects (the visible marks) involved must also be divested of any geometrical structure consistent with classical physics. Either individually or collectively, these traces are no longer seen as arising from any processes allowing for a classical-like physical description.

Heisenberg's first paper on quantum mechanics appears to reflect the situation just described. It suspends the application of classical physics to the emergence of quantum data and the very possibility of configuring these data and the processes responsible for them accordingly. Instead, it treats them as certain effects divorced from all forms of classical emergence or history. His introductory elaborations in the paper itself suggest nearly as much, as Bohr's commentary on the paper, cited above, makes clear. Heisenberg proceeded from the disassemblage of the experimental data (in this case, spectra) in question as possible or previously presumed effects of classical radiational processes to relating these data to infinite square tables, matrices, of complex-number quantities. These quantities are linked to the probabilities of transitions of an atom, or an electron in this atom, from one stationary state to another by emitting quanta of radiation. It is this new arrangement that enabled him and others (this took a few months) to formalize the predictions concerning these effects of the interaction between quantum objects and measuring instruments upon in a new way. The proper manipulation of these tables became quantum mechanics, in its matrix version (Schrödinger's equation came later). As explained earlier, the new (as opposed to classical physics) kinematics was defined by the matrix elements linked to the probabilities of transitions from one physical state (of an electron in an atom) to another, arranged into these tables, now used as variables of equations borrowed from classical mechanics and applied there to standard physical variables.

As a result, two new major physical and philosophical concepts are introduced. The first is a new concept of a physical variable (new kinematics) and, at least in principle (the specifics were developed later in the work of, in addition to Heisenberg, Born and Jordan) a new matrix and eventually, with Dirac and von Neumann, the Hilbert-space formalism of quantum mechanics. The second, again, at least in principle, is a new concept of probability in physics, the probability that defines, as against the causality of classical mechanics or, at the ultimate level, classical statistical physics, even the predictions concerning primitive individual events. This concept was developed, in physical terms, by Born in his probabilistic interpretation of the wave function and in the epistemological terms by Bohr, as considered in Chapter 4.

Hegel would have said that Heisenberg released the phenomena in question from their *immediate* relationships to classical physical concepts, or he established mediation (*Vermittlung*) of these phenomena by these concepts as applicable to quantum objects (as responsible for these phenomena). Or, rather, since these relationships appeared to be no longer made possible by nature itself, he abandoned any attempt to re-establish this (classical) type of mediation. Instead, he proceeded to the invention of new concepts, at least mathematical-physical concepts, and a very different (ultimately, epistemologically nonclassical) type of mediation between these phenomena and his new quantum mechanics. As we have seen, in the process he also established an entirely new kind of relationship between mathematics and physics.

Heisenberg's achievement is pretty miraculous by any criteria, and the significance of his discovery was quickly grasped by Heisenberg's contemporaries, with much relief by some, especially by Bohr, and with a considerable degree of discontent by others, especially by Einstein. Einstein of course recognized the practical physical significance of Heisenberg's discovery, even though he clearly preferred Schrödinger's classical-like approach. Although, as indicated in Chapter 1, the process of Heisenberg's discovery was more protracted and complex (and thus also less dramatic or miraculous than presented here), it would appear that at least at some point, however briefly, the type of situation just described must have been confronted by Heisenberg. It became part of the confrontation between his thought and quantum phenomena, to which it appeared impossible to give the kind of order that classical physics and its thought were able to give to classical phenomena. As it happened, quantum phenomena required and ultimately received a very different form of theoretical order, the order that is epistemologically nonclassical in nature. I shall return to this question of the relationships between thought and chaos in physics and elsewhere later in this chapter. For the moment, however that may have happened, Heisenberg's invention, too, was "the highest form of musicality in the sphere of thought," as Einstein famously described Bohr's semi-classical theory of the hydrogen atom of 1913. He said: "That this insecure and contradictory foundation was sufficient to enable a man of Bohr's unique instinct and tact to discover the major laws of the spectral lines and of the electron shells of the atom together with their significance for chemistry appeared to me like a miracle—and appears to me as a miracle even today [in 1949]. This is the highest musicality in the sphere of thought" (Schilpp 1949, pp. 45-47).

It also appears, however, that one needs to be at least as good as Bohr or Heisenberg, or in spite of himself, Einstein, to enable physics to remain physics, and perhaps to become physics more than it has even been classically, under these nonclassical conditions. This task may indeed require "the highest musicality in the sphere of thought." It may even require the highest musicality in the highest sphere of thought, where musicality and thought become one and the same. But then perhaps this is what true *thought* is, born in our confrontation with chaos, such as the one with which quantum phenomena threatened our thinking, accustomed to the forms of order created by classical physics or classical philosophical and epistemological thought, and to which we were able to give order by quantum mechanics. But then, again, the true, highest thought and its highest musicality may need chaos to arise.

Heisenberg was of course far from being done with his major contributions (less miraculous as they might have been). His work with Born and Jordan on matrix mechanics and the uncertainty relations or his work with Pauli on quantum field theory was still ahead of him, to list his most momentous achievements. To one degree or another, all this work was shaped by his capacity to detach the phenomena in question from the old, no longer workable, ways of thinking and to invent new ways of thinking in order to understand these phenomena or at least to work with them and to move physics forward. It is this capacity that was decisive in his discovery of quantum mechanics, and one finds the same capacity in Bohr's 1913 theory of the hydrogen atom, admired by Einstein for "the highest musicality in the sphere of thought" found there. This highest musicality in the sphere of thought is equally found in Heisenberg's work on quantum mechanics, or of Einstein's own work on the old quantum theory and then relativity, special and general theories alike. Einstein's work and thought have their Bohrian and Heisenbergian aspects as well, especially manifest in Einstein's capacity to depart from the old ways of thinking about physical phenomena and to invent new, radically new ones.

Heisenberg's thought had also a few things to offer to physics and philosophy, or to thought itself, along other lines, those that proceeded more along the Kantian lines of critical philosophy, the critique of concepts, physical and philosophical. Heisenberg's (1929) Chicago lectures were significantly influenced by and paid a tribute to Bohr's preceding work on complementarity as physics and as epistemology, although Heisenberg in fact manages to avoid some of the problems of Bohr's Como version even at this point, in part through his more Kantian critical approach. I shall comment on this influence and its significance for Heisenberg's critical or, one might say, "deconstructive" argument presently. (Schrödinger's 1926 wave quantum mechanics played a role as well.) Some (not all!) of the key features of Heisenberg's argument itself are, however, closer to Bohr's later (post-EPR) work on complementarity, as considered in this study. This is not surprising, since, as discussed in Chapter 1, Heisenberg's own earlier work (before Schrödinger's equation and Bohr's Como version of complementarity) on quantum mechanics is closer to Bohr's later views, in turn indebted or proximate to Heisenberg's early work (it is difficult to disentangle or unambiguously sequence influences here).

This proximity is especially apparent in Heisenberg's extraordinary appendix (expanded for the English publication of the lectures), "The Mathematical Apparatus of the Quantum Theory." This appendix is much more than an appendix and much more than its title would suggest, unless we understand, as Heisenberg perhaps did, this apparatus as irreducibly linked to physics, specifically the new kinematics of quantum theory in terms of manifest physical effects of measurement, as discussed earlier in this study. Heisenberg's philosophical attitude in the lectures is defined primarily by that part of Bohr's philosophy, on which Heisenberg was later to comment in the statement cited earlier: "Bohr was primarily a philosopher, not a physicist, but he understood that natural philosophy in our day and age carries weight only if its every detail can be subjected to the inexorable test of experiment." Heisenberg's own argument—indeed, his lectures also offer us certain "*philosophical* principles of quantum theory"—proceeds in philosophical terms of the *critique* of concepts, closer to the Kantian sense of critical philosophy, and in effect establishes a certain Kantian axis in the epistemology of modern physics, and to some degree even in physics itself. Heisenberg starts with special relativity as an experimentally grounded, but, in general, more broadly, including philosophically, based, critique of classical concepts. He writes:

> Thus it was characteristic of the special theory of relativity that the concepts "measuring rod" and "clock" were subject to searching criticism in the light of experiment; it appeared that these ordinary concepts involved the tacit assumption that there exist (in principle, at least) signals that are propagated with an infinite velocity. When it became evident that such signals were not to be found in nature, the task of eliminating this tacit assumption from all logical deductions was undertaken, with the result that a consistent interpretation was found for facts that had seemed irreconcilable. A much more radical departure from the classical conception of the world was brought about by the general theory of relativity, in which only the concept of coincidence in space-time was accepted uncritically. According to this theory, ordinary language (i.e. classical [physical] concepts) is applicable only to the description of experiments in which both the gravitational constant and the reciprocal of the velocity of light may be regarded as negligibly small. (Heisenberg 1930, p. 2; translation slightly modified)

Heisenberg's parenthesis is crucial in indicating some of the key complexities of the situation, as considered earlier: first, those concerning the relationship between ordinary (everyday) language and classical physics, including as ultimately inapplicable to the ultimate objects of nonclassical physics; and second, those concerning the unknowable and inconceivable nonclassical efficacious dynamics of certain observable effects. The latter point becomes especially crucial once Heisenberg moves to quantum mechanics. This critique may also be seen in (parallel) terms of Bohr's nonclassical epistemology of phenomena as effects of quantum-mechanical efficacious processes, although the latter (deliberately followed or de facto enacted, as it nearly was in Heisenberg's initial discovery of quantum mechanics) allows one to proceed otherwise than by means of the type of critique Heisenberg has in mind here. For the moment, this critique extends and radicalizes this Einsteinian scheme to a nonclassical level, or just about. Heisenberg develops his critique in the next two chapters, "Critique of the

Physical Concepts of the Corpuscular Theory of Matter" and "Critique of the Physical Concepts of the Wave Theory." Heisenberg's analysis is extraordinary and it is regrettable that I have to bypass the details here. I do so, however, primarily because the physics and (nonclassical) epistemology, and physical and philosophical conceptuality that he ultimately arrives at have already been established in this study. On the other hand, I shall address the essential aspects of the two critiques in question and argue that at stake in them, as critiques, is a Kantian critique of the classical concepts and theories of particles and waves, and their motion. According to Heisenberg:

> In the foregoing chapter ["Critique of the Physical Concepts of the Corpuscular Theory of Matter"] the simplest concepts of the wave theory, which are well established by experiment, were assumed without question to be "correct." They were taken as the basis of a critique of the corpuscular picture, and it appeared that this picture is only applicable within certain limits, which were determined. The wave theory, as well, is only applicable with certain limitations, which will now be determined. Just as in the case of particles the limitations of a wave representation were not originally taken into account, so that historically we first encounter attempts to develop *three-dimensional* wave theories that could be readily visualized (Maxwell and de Broglie waves). For these theories the term "classical wave theories" will be used; they are related to the quantum theory of waves in the same way as classical mechanics [of particles] to quantum mechanics. [...] (The reader must be warned against an unwarrantable confusion of classical wave theory with the Schrödinger's [quantum] theory of waves in a phase space.) After a critique of the wave theory concept has been added to that of the particle concept all contradictions between the two disappear—provided only that due regard is paid to the limit of applicability of the two pictures. (Heisenberg 1930, p. 47)

Heisenberg, thus, introduces a radical critical program, which is Kantian in its origin but extends beyond Kant and is closer to Nietzsche and those, such as Heidegger, who follow him (or quantum-mechanical theorists such as Bohr and Heisenberg, to whom Heidegger specifically refers as well). First, it is a Kantian critique in more general terms of exploring the limits and establishing new boundaries, de-limitations, of the concepts, here, those of classical physics, from which we inherit these concepts. We may need to retain these concepts within certain limits (e.g. as applicable to measuring instruments) put to a new use (e.g., to apply in a more limited fashion and perhaps only provisionally to quantum objects). Secondly, more innovatively, Heisenberg here employs a particular form of such a critique as a *mutual* critique of each theory by means of the others. He uses conflicting classical theories, that of wave and that of particle, against each other in order to properly re-delimit them and, by so doing, to establish an epistemological and conceptual architecture of quantum mechanics, where such concepts as wave and particle could still be used within these new limits. By the same token, a certain interpretation of quantum mechanics is established as well. This interpretation largely (albeit not altogether) follows Bohr's complementarity, established by that time in its Como version, to which Heisenberg expressly refers (Heisenberg 1930, pp. 62-65). In particular, as discussed in this study, the approach allows one to avoid contradictions that plagued and were perhaps unavoidable in quantum mechanics, in particular by virtue of an uncritical use of such concepts as wave and particle (i.e. by simply borrowing them from classical physics, without a proper analysis and de-limitation of their applicability). Hence, his conclusion: "After a critique of the wave theory concept has been added to that of the particle concept all contradictions between the two disappear—provided only that due regard is paid to the limit of applicability of the two pictures."

Heisenberg's overall approach in the Chicago lectures is different from that of Bohr, either in the Como lecture or in Bohr's post-EPR arguments, which radicalize the epistemology of complementarity so as to reach its nonclassical limits. Or, more accurately, it is centered differently, namely, on a critique and re-delimiting of classical concepts rather than on the invention of new concepts, as in Bohr's work or in Heisenberg's own earlier work. This different centering or focusing is in part due to the fact that Heisenberg is already in possession of Bohr's concepts, or those of his own or those introduced by Dirac, Schrödinger, and other founders of quantum mechanics, which he re-derives via this critique. On the other hand, Bohr's own invention and presentation of his concepts, say, those in the Como lecture, involved a critique of classical concepts, as does Heisenberg's work on his matrix mechanics or, even more so (and closer to the Chicago lectures), on the uncertainty relations. It is a question of relative balance of the invention of concepts and the critique of concepts (or of course of an epistemological analysis) in nonclassical epistemology.

It should not be forgotten either that Heisenberg's invention of his new mathematical-physical concepts (a reinvention of the mathematics of matrices and an invention of the corresponding physical concepts) made the whole scheme possible in the first place. Indeed, as discussed earlier, these concepts are also philosophical insofar as they involve a reconceptualization of our organization, architecture, of the relationships between mathematical (matrix) elements and the physical data in question in quantum mechanics, and moreover, doing so strictly, irreducibly in terms of probabilistic estimates. As was also discussed earlier, Heisenberg's discovery was, more generally, governed by an extraordinary and unprecedented (although Bohr's earlier thinking in his 1913 theory of the hydrogen atom comes close) epistemological-philosophical vision.

On the other hand, it may be argued that Heisenberg's more critical approach in the Chicago lectures allows him to avoid some of the problems of Bohr's Como argument. In a way, Heisenberg's lectures may be seen as refining Bohr's Como argument, a task that Bohr himself accomplishes later rather differently and, again, closer to Heisenberg's argument in his earlier work. Here, however, Heisenberg proceeds through a critique of classical concepts. Heisenberg's last qualification in the passage under discussion (to the effect that the avoidance of contradictions between the wave and the particle pictures is only possible if due regard is paid to the limit of applicability of the two pictures) is, again, especially crucial here. Earlier in his analysis, Heisenberg makes a decisive, physically and philosophically, point (ultimately correlative to the uncertainty relations). He says:

> As a matter of fact, it is experimentally certain only that light sometimes behaves as if it possessed some of the attributes of a particle, but there is no experiment which proves that it possesses all the properties of a particle; similar statements hold for matter and wave motion. The solution of the difficulty is that the two mental pictures which experiments lead us to form—the one of particles, the other of waves—are both incomplete and have only the validity of analogies which are accurate only in limiting cases [where quantum and classical physics give the same results]. (Heisenberg 1930, p. 10)

This is why what is in question in Heisenberg's critique is, first, an investigation of the proper limits of classical concepts and theories under critique, and, second, their workings against each other within and in order to establish a nonclassical framework, which the phenomena in question in quantum mechanics may require. Eventually, in the wake of the EPR argument, Bohr arrived at a more radical interpretation, which is much closer to Heisenberg's initial work on quantum mechanics. This interpretation does not depend on either wave or particle theories or properties, not even in partial terms (as indicated by Heisenberg's statement above) in describing quantum objects, to whose characterization and behavior neither, or indeed any, description or theory is applicable. At most some properties of either theory are retained at the level of the effects of the quantum (and hence ultimately in turn indescribable) interaction between quantum objects and measuring instruments upon the latter. Nor, accordingly, would one need to depend on a mutual critique of the type just described in developing this interpretation. One might proceed in this way, too, as Heisenberg does in his Chicago lectures. But one need not, as is in turn shown by Heisenberg's discovery of quantum mechanics or his collaboration on matrix mechanics with Born and Jordan, or in their own work or that of Dirac, in extending and properly developing Heisenberg's discovery, the greatest physical achievement here. Philosophy, but some physics too, at least a proper understanding of the physical content, a physical interpretation, of the theory, had to wait (not very long) for Bohr.

5. "THE BASIC PRINCIPLES OF SCIENCE": NONCLASSICAL EPISTEMOLOGY, SCIENTIFIC DISCIPLINARITY, AND THE PHILOSOPHY OF PHYSICS

Quantum mechanics offers us, to return to Bohr's formulation, a tremendous "opportunity of testing the foundation and scope of some of our most elementary concepts." As we have seen in the case of quantum field theory in Chapter 5, we are far from finished with this nonclassical testing and perhaps new, more severe and more unimaginable, tests and trials are in the offing. In physics, while some of the currently available theories, such as much of classical physics or (nonrelativistic) quantum mechanics, may be seen as complete within their limits, overall our fundamental physics and hence our knowledge concerning the ultimate constititution of the physical world remain manifestly incomplete. We have classical physics, relativity, quantum mechanics, quantum electrodynamics and quantum field theories, or still other theories, such as string theories, by now extended into "branes" theories; and each of these areas appears to branch out nearly endlessly and uncontainably. These theories describe various macro and micro aspects of the physical world and of our interaction with it by means of experimental technology. They do so sometimes in classical-like ways, sometimes in quantum-mechanical-like ways, sometimes by combining both. The epistemological status of many of these theories is far from established or even addressed, and some of them are highly speculative physically. Quantum mechanics (at least as complementarity) and, by implication, its extensions rigorously suspend the possibility of physical description at the level of the ultimate constituents of matter themselves. This may or may not continue to be the case, once new theories or new interpretations of them (or of existing theories, such as quantum mechanics) take shape. As also discussed in Chapter 5, still more radical epistemological configurations are not inconceivable either. For the moment we can at best correlate some among the available physical descriptions and try to maintain their consistency with experimental data within sufficiently workable limits. (Some of these theories are manifestly inconsistent with each other.) It remains an open question whether physics can ever be reasonably brought together, or even needs to be. It is conceivable that, as Einstein hoped, future theories, or new data, will transform physics, and will do so by means of a more homogeneous single theory, or at least a set of more homogeneous theories, which will, in particular, be no longer quantum-mechanical-like or epistemologically nonclassical. It is also conceivable that, as Bohr thought, future developments will preserve the joint significance of classical-like and quantum-like physical, or philosophical, theories in describing, or making it impossible to describe, the ultimate constituents of matter. This conjunction of classical and nonclassical theories has defined the century of physics that began with Planck's discovery of the quantum of action in 1900. It is also possible that the future will produce not only as yet unencountered but as yet inconceivable configurations of nature and science, or of matter and mind, assuming that these, for now seemingly inescapable, concepts will be retained.

It would be difficult to imagine anything of a lesser complexity in mathematics, biology, or information and computer sciences, as all these fields invade and fuse with each other ever more aggressively, without, it appears, worrying too much about interdisciplinary hurdles. (Mathematics and science do not appear to have ever existed apart from interdisciplinary projects.) Will these developments give an ever-increasing role to nonclassical epistemology? The answers, it appears, would be the same as the one just given for physics. We do not know at the moment, but the significance and even expansion of the nonclassical character of our knowledge in these fields is at least as likely as any return to classical knowledge at the ultimate level. Classical knowledge, once again, retains its significance both in nonclassical knowledge and in its own right in many areas of modern mathematics and science or, of course, elsewhere.

According to Bohr, the nonclassical epistemology of quantum mechanics as complementarity is not only compatible with the basic principles of science, but provides room for new physical laws, the discovery of which is the most fundamental task of all science. A few circumstances of quantum measurements or phenomena (now also in Bohr's sense) and of quantum mechanics itself in Bohr's interpretation of them as complementarity, as considered in this study, are especially pertinent to his argument and are worth briefly revisiting here. Certain parts of measuring instruments and the observable effects of their interaction with quantum objects are described classically (and thus also in the realist way), although the sum total of these effects cannot be accounted for by means of classical physics, and requires quantum theory. The ultimate nature of this interaction is quantum, however, which makes it in practice uncontrollable (thus disabling the simultaneous exact measurement of both conjugate variables) and, in its quantum aspects, theoretically indescribable. In this latter respect it is no different from any quantum process, which, in Bohr's interpretation, is never theoretically describable as such; only its effects (upon measuring instruments) are, as Bohr says in the statement just cited. "Quantum effects" are all that is available to us, never quantum causes. It is the irreducible interaction between quantum objects and measuring instruments that is responsible for the radical epistemology of quantum effects without quantum causes, or without any ultimate causes. The interaction in question cannot be seen as the ultimate cause here, given that its ultimate nature is itself quantum, and hence beyond any conception, that of causality included.

As explained earlier in this study, Bohr stresses this interaction and its irreducible nature throughout his writing, and he argues that this interaction defines any phenomena that can be meaningfully considered in quantum physics. These are the circumstances that he has in mind when he says that "under these circumstances an essential element of ambiguity is involved in ascribing [any] conventional physical attributes [single or joint] to quantum objects [themselves]" (*PWNB* 2, p. 40). Given that this interaction is irreducible, he is able to argue that "a criterion [of reality] like that proposed by [EPR] contains [...] an essential ambiguity when it is applied to the actual problems with which we are here concerned" (*PWNB* 4, p. 75). For, in view of this interaction, we cannot unambiguously ascribe, as EPR, again, do in accordance with their criterion, independent properties to quantum objects, ultimately even to a single such property, let alone both complementary ones, or again, at the limit, even independent identity to quantum objects.[69] Thus, "the apparent contradiction [found by EPR] in fact discloses only an essential inadequacy of the customary viewpoint of natural philosophy for a rational account of physical phenomena with which we are concerned in quantum mechanics" (*PWNB* 4, pp. 74-75). Instead, the irreducibility of this interaction "entails the necessity of the final renunciation of the classical ideal of causality and a radical revision of our attitude towards the problem of physical reality," or, again, at least this interpretation of quantum mechanics allows one to effectively reply to the EPR argument. This interpretation of quantum mechanics rigorously limits and re-delimits both causality and reality in the sense of classical physics and of the classical philosophy of nature. This, again, does not mean that these concepts are now abandoned altogether, since beyond their significance in classical physics itself, which we still use, we still use these concepts and indeed classical physics for our description of measuring instruments.

I have considered the physical and philosophical aspects of Bohr's counterargument to EPR in detail earlier in this study, and there is no need to further consider these aspects here. On the other hand, Bohr's language in the passage just cited and throughout his works manifests his concern with the disciplinary nature and role of physics as an experimental-mathematical science of nature. This aspect of his thought and writing is worth further considerations, which I would like to offer here, both in their own right and as exemplifying my argument that nonclassical epistemology, in physics or elsewhere, has as much rigor, richness, and disciplinary as do classical theories, such as classical physics. In addition, this argument extends and deepens our view of what the philosophy of science is or could be, and indeed suggests a new form of philosophy of science. To cite Bohr's conclusion:

> [...] the argument of [EPR] does not justify their conclusion that quantum mechanical description [sic] is essentially incomplete. On the contrary this description, as appears from the preceding discussion [i.e., in Bohr's interpretation], may be characterized as a rational utilization of all possibilities of unambiguous interpretation of measurements, compatible with the finite and

---

[69] As I explained earlier, the nonlocality part of the EPR-type argument can be readjusted so as to refer only to the outcomes of measurements, rather than to quantum objects. These considerations would not affect either Bohr's interpretation (including its local character) or the argument presented here.

> uncontrollable interaction between the [quantum] objects and the measuring instruments in the field of quantum theory. In fact, it is only the mutual exclusion [in view of this interaction] of any two experimental procedures, permitting the unambiguous definition of complementary physical quantities [such as position and momentum], which provides room for new physical laws [i.e., the laws of quantum mechanics], the coexistence of which might at first sight appear irreconcilable with the basic principles of science [but is ultimately not]. It is just this entirely new situation as regards the description of physical phenomena that the notion of *complementarity* [now in Bohr's extended sense] aims at characterizing. (*PWNB* 4, p. 80).

As noted earlier, Bohr's syntax is cumbersome and even tortured here, which accounts for my interpolations, which, I hope, help, although I am hesitant to intervene too much into Bohr's writing. It is clear, however, that Bohr's first sentence omits "reality" here, as against EPR's formulation of the problem. This omission may have been deliberate, and is certainly telling. The completeness of quantum-mechanical physical description may no longer allow for reality in EPR's sense. That is, it may no longer allow for an independent physical reality, defined by postulating the existence, on the classical model, of physical properties of objects (or, again, conceivably, such classical-like objects themselves) as independent of their interaction with measuring instruments. Instead, as discussed throughout this study, in Bohr's interpretation, quantum-mechanical physical description refers to "phenomena" defined as the overall experimental arrangements within which quantum effects, marks left in our measuring devices, manifest themselves.[70] Thus, according to Bohr the irreducible interaction between quantum objects and measuring instruments, while indeed incompatible with the

[70] That, again, is not to say that "the quantum world" or the corresponding (ultimate?) level of the constitution of matter does not exist, but that the attribution of physical properties, including those of individual identity of particles, or conversely of wave-like substances, may not be possible at that level. Nor, however, would it follow (as some contend) that this suspension of the independently attributable particle identities, such as those of two "particles" in the EPR situation, in fact entails nonlocality. Two quantum entities involved would still be spatially separate, and, according to Bohr, there is in the EPR case certainly "no question of a mechanical [i.e. physical] disturbance of [one] system under investigation" by our interference with the other quantum system involved in the EPR thought experiment (*PWNB* 4, p. 80). It is just that we cannot attribute independent physical properties, ultimately, even that of a "particle," to them. Once we assume that we can, as Einstein did, nonlocality indeed appears to follow. Thus, as discussed earlier, his argument is not logically wrong. His assumptions may well be wrong, or at least are not warranted by the experimental evidence in question.

classical ideals of causality and reality, is by no means incompatible with "the basic principles of science." The compatibility between them, however, is only possible if one properly interprets what is in fact available to an unambiguous account in the entirely "new situation" we encounter in the field of quantum theory and what this theory actually unambiguously accounts for and how it accounts for what it accounts for.[71] Accordingly, Bohr's interpretation of quantum mechanics as complementarity, which rigorously follows this requirement, "may be characterized as a rational utilization of all possibilities of unambiguous interpretation of measurements, compatible with the finite and uncontrollable interaction between the [quantum] objects and the measuring instruments in the field of quantum theory." This interaction also entails "the *impossibility of any sharp separation between the behavior of atomic objects and the interaction with the measuring instruments which serve to define the conditions under which the phenomena appear*" (Bohr 1949, *PWNB* 2, pp. 39-40; Bohr's emphasis), which, arguably, defines most the epistemological basis of complementarity. In the process, complementarity also "provide[s] room for new physical laws [the laws of quantum physics], the coexistence of which might at first sight appear irreconcilable with the basic principles of science."

Bohr's argument is, thus, as follows: were it not for these new epistemological conditions defined by "the *impossibility of any sharp separation between the behavior of atomic objects and the interaction with the measuring instruments which serve to define the conditions under which the phenomena appear,*"

a) EPR would be right: quantum theory would be incomplete, or else nonlocal (or at least short of an interpretation that ensures both completeness and locality); and
b) More strongly, there would be no room for the laws of quantum mechanics as physical laws (the same type of parenthesis is required).

The laws of quantum mechanics may appear, in particular to EPR, to be "irreconcilable with the basic principles of science." Quantum mechanics, however, accounts for its data as well as any classical physics does for its data, which is what EPR tried, ultimately unsuccessfully, to question. It may, thus, depend on which principles one sees as basic to science, both in general and insofar as such principles can be applied

[71] As discussed earlier, the essential ambiguity of the EPR criteria, as applied in quantum mechanics, arises from their failure to do so, however subtle and revealing of new aspects and "mysteries" of the quantum it may be. In particular, it is, again, the failure to see that "an essential element of ambiguity is in ascribing conventional physical attributes to [quantum] objects themselves," ultimately single (rather than only complementary) such attributes, or, again, even seeing such objects as particles. This is the same ambiguity.

in the case of quantum physics, as a mathematical science of physical objects and phenomena. The distinction between quantum objects, which are unobservable or even inconceivable and hence can never be phenomenalized, and observable phenomena, which are manifest in measuring instruments, is a crucial part of both his interpretation of quantum phenomena and his argument on the particular point in question at the moment. Bohr argues as follows. As against the necessity of conforming to a particular (philosophical) criterion of physical reality, demanded by Einstein, Bohr sees the following principles as most basic to science. The first is the logical consistency of a given theory; the second is the correspondence between the predictions provided by the theory and the available experimental data; and the third is the capacity of the theory to "[exhaust] the possibilities of observation," in other words, its completeness. If one sees these principles as basic, then Einstein's argumentation could not be directed towards demonstrating the inadequacy of quantum mechanics, since, Bohr argues, quantum mechanics clearly conforms to these principles within its proper scope (*PWNB* 2, pp. 56-57).[72]

Accordingly, along with the reexamination of the classical ideas and ideals of reality and causality necessitated by quantum mechanics, a similar, and indeed parallel and interactive, reexamination of what constitutes the basic principles of science appears to be rigorously necessary. For, on the one hand, the EPR criterion of reality cannot unambiguously apply to the "entirely new situation as regards the description of physical phenomena" in question in quantum mechanics. On the other hand, quantum mechanics itself comprehensively accounts for these phenomena and, hence, is as rigorously scientific as any (classical) mathematical science in every respect (other than causality and reality). Clearly, one needs to (re)consider what the basic principles of science are. This is a crucial point, stressed throughout this study: physics itself, not philosophy, requires or at least may require this reconsideration, since, as Heisenberg observes in the passage cited earlier, "natural philosophy in our day and age carries weight only if its every detail can be subjected to the inexorable test of experiment," as, according to Heisenberg, Bohr deeply understood. In the present case this test may entail the irreducibly nonclassical character of natural philosophy once one considers nature at the quantum level, the level of its ultimate constituents. On the other hand, as a result of this reconsideration, physics may and does impact philosophy, even, as I argue here, beyond the philosophy of nature or the philosophy of physics. The basic principles of science must be weighed and, if necessary, adjusted accordingly.

As understood by Bohr, the basic principles of science are in accord with the main defining aspects of the project and practice of classical physics, beginning with Galileo, to whom Bohr specifically refers in this context (*PWNB* 3, p. 1). It is true that

---

[72] While the above qualifications, especially those concerning nonlocality, must be kept in mind, they would not affect the points made at the moment.

Einstein and many others would see certain other principles as equally basic, which principles are, I argue here, more philosophical than physical, although, I also argue, this distinction is not absolute and cannot be established unconditionally, once and for all. Accordingly, Bohr's argument concerning quantum mechanics also suggests that the basic principles of science qua science may, at a certain point, come into conflict with those philosophical principles, however consistent the latter may be with classical physics. This conflict leads to a different understanding of the character of science qua science, to a different philosophy of science, along with and as a result of a different philosophy of knowledge (epistemology) and nature.

What are the basic principles of science, according to Bohr's view? What would define the science and the discipline (in either sense) of physics as, to return to Galileo's language, a (modern) "mathematical science of nature"? There are, as I see it, more or less four such principles, which I indicated above and which I shall now formulate more rigorously. Further qualifications and nuances are necessary, in part in view of the massive recent reconsideration of the nature and the character of scientific knowledge. This reconsideration arises from the work of such authors as Thomas Kuhn, Imre Lakatos, Paul Feyerabend, and their followers, more recently those working in the so-called constructivist science studies, such as Bruno Latour or David Bloor and his school. It may be shown that, in essence, these formulations are consistent with this reconsideration—at its best. The works involved are not without their own problems, sometimes as severe as those of the classical views in question in this reconsideration, with which, at their best, the principles in question can in turn be correlated. Of course, a broader and longer history of history and philosophy of science, at the very least in the twentieth century, is relevant here, including as concerns the main predecessors of the figures just mentioned, such as Karl Popper or Rudolph Carnap. I have addressed some of these developments and debates on earlier occasions (Plotnitsky 2002; also Plotnitsky and Read 2001). The argument offered at the moment in part derives from this earlier analysis, including as concerns the main point I would like to make, although it also departs from this analysis in several significant respects. This point is the compatibility of nonclassical epistemology and the kind of philosophy of nature, knowledge, and physics it implies, which are my main concerns here, even with a reasonably traditional view of the basic principles of science, at least some among these principles, but, arguably, these are also the most crucial ones. Indeed, further complexities just mentioned appear to move us even further toward nonclassical thinking concerning nature, knowledge, and science, while, by contrast, posing greater problems for a classical view of all three. The principles themselves in question can be described as follows.

1) *The mathematical character of modern physics*. I think, with both Galileo and Bohr, or Newton or Einstein, that this mathematical character is what defines modern physics most fundamentally. I understand by this character the usage of mathematics as a

particular way of offering *convincing* arguments about certain aspects of and certain facts pertaining to the physical world, rather than only a mathematical representation of the ultimate nature or structure of this world.[73] The latter is rigorously impossible to do in quantum physics, at least in Bohr's interpretation. This point is crucially implicated in the Bohr-Einstein debate, and there also appear to be significant differences in this respect between Galileo's and Newton's project and philosophy of science, or of nature itself. For Galileo a science of motion is a construction of convincing mathematical arguments about certain facts and aspects of nature. For Newton, it is more (there are further complexities to Newton's thought on the subject as well) a representation of nature, grounded in the classical realist claim that nature possesses a structure that can ultimately (at least by God) be represented mathematically. The latter view in fact defines the attitude of most physicists. Exceptions are few.[74]

2) *The principle of consistency*. These theories must offer logically consistent arguments, even though they may and perhaps, especially in quantum mechanics, must exceed their mathematical aspects and involve verbal and ordinary-language components, as Bohr notes throughout his works. In his *The Interpretation of Quantum Mechanics*, Roland Omnès bases his whole interpretation of quantum mechanics around logical consistency (Omnès 1994, Omnès 1999). There are important further nuances to this principle as well, which I must bypass here. In any event, these theories must be as logically consistent as anything can be.

3) *The principle of unambiguous communication*. These theories, in their mathematical and nonmathematical aspects alike, must allow, within the practical limits of the functioning of science, for the (sufficiently) unambiguous communication of both the experimental results and theoretical findings involved, including, again, those involving statements in ordinary language. This is also what Bohr's interpretation of quantum mechanics as complementarity "provides room for," in part by virtue of exploring the possibilities of the unambiguous definition of all physical variables and aspects of physical description involved, as against ambiguities arising if one adopts a classical epistemological position. As we have seen, the concepts of "unambiguous definition" and "unambiguous communication" become especially crucial for Bohr in the

---

[73] An appeal to "convincing arguments" entails additional complexities, such as those involved in the recent developments and debates, mentioned above, concerning philosophy, history, and sociology of science, and I cannot address these complexities in detail here. However, the *core* of my argument concerning the disciplinary workings of physics under the nonclassical epistemological conditions would remain in place, and, I would argue, is even more firmly established once these complexities are taken into account. The argument itself will of course become more nuanced.

[74] For a further discussion of this difference between Galileo and Newton, see (Plotnitsky and Reed 2001).

wake of the EPR argument, where the question of the unambiguous definition of these variables becomes linked to the question of the locality of quantum mechanics.

4) *The principle of experimental rigor* (based, at least from Galileo on, on the concept of measurable quantities). Our physical theories must correspond to and, within their limits, exhaust the experimental data they aim to account for, although this data is of course itself subject to interpretation. Certainly, in quantum physics the question of how one interprets its data is as crucial as, and reciprocal with, that of how one interprets quantum theory. Much more is to be said on this point as well, even leaving aside the question of the social construction of theories and related arguments, which would affect the principles of consistency and unambiguous communication as well. The principle itself, however, remains crucial.

Quantum mechanics as complementarity is the first theory ever that at the very least allows for, even if not necessitates (it may ultimately do this too), for the compatibility between "the basic principles of science" and nonclassical epistemology. This is a crucial point, and, while my invocation of Galileo in the above outline of these principles is aimed to emphasize the disciplinary continuity of all modern physics as a mathematical science of nature or, to speak closer to Galileo, a set of such sciences, it cannot diminish the epistemological revolution enacted by quantum mechanics as complementarity. The Galilean and Bohrian views of physics are not identical either, and the differences between them are significant, historically and conceptually. Galileo's view is ultimately classical, even though it is closer to a nonclassical view than is a Newtonian view of physics. It may, accordingly, be worthwhile to reflect, with the help of the preceding analysis, on this difference between Galileo and Bohr further here in other to make the richness, depth, and significance of the quantum-mechanical epistemological revolution more apparent.

Let us recall first that, as a nonclassical theory, Bohr's complementarity, at least, again, in its post-EPR version, strictly prohibits, "*in principle* exclude[s]," any attribution, or even any conception, concerning the physical properties of the ultimate objects and processes in question in quantum mechanics, quantum objects and processes. These objects, however, are now considered as *objects* and *processes* of quantum mechanics as a *particular physical theory* and, moreover, in a *particular interpretation* of this theory as complementarity. In other words, these objects are *idealized* in this way by quantum mechanics as complementarity. On the one hand, this view implies that the ultimate constitution of nature responsible for the experimental data in question may not conform even to this idealization—that is, it may be inaccessible even as inaccessible, unknowable even as unknowable, inconceivable even as inconceivable, and so forth. On the other hand, this view also leaves room for alternative, possibly including classical-like interpretations of quantum mechanics or for alternative theories of these data, and hence possibly of nature itself. I shall, again, put aside such alternatives, such as more realist interpretations of quantum mechanics or (these are, again, different theories) various versions of Bohmian mechanics, since their viability is a complex matter and my main concern for the moment is the epistemological and philosophical nature of complementarity or, more generally, nonclassical theorizing. Nonclassically, the situation is as follows.

Something irreducibly inaccessible (unrepresentable, unknowable, inconceivable, and so forth), serving as a theoretical idealization links us, without any conceivable representation or any conceptual relation, to yet something else that is equally and possibly even more remotely inaccessible. This link, however, allows this idealization to make, by means of the mathematical machinery of quantum mechanics, correct, albeit generally statistical, predictions concerning the outcome of the experiments in question on the basis of the data obtained in other already performed experiments. The nature of this link is irreducibly enigmatic, mysterious insofar as, in contrast to classical mechanics, the theory (quantum mechanics as complementarity) offers no physical description, however idealized, or arguments concerning the nature of quantum objects and processes. Nor does it or possibly nature itself, physical or human, offer us any reasons for understanding, or the possibility to imagine, how this is possible. In this view, quantum mechanics becomes no longer physics in its conventional sense, that is, in the sense physics was understood previously, but a form of information processing between measuring instruments and human beings who use them (a view, again, adopted in the so-called quantum information theory). As was also discussed earlier, however, this does not mean (along more positivistic lines of thought) that complementarity is not concerned with quantum objects and processes, but, again, only that it cannot and even disallows one to say anything about them or even conceive of them. It is as unknowable and inconceivable that they are responsible for the phenomena (now defined in Bohr's sense of the effects of the interaction between quantum objects and measuring instruments) that we can predict and study.

This (potentially unavoidable) "absence" of physics in the quantum-mechanical description of nature, which suspends any description or even conception of the ultimate constitution of nature even at the level of idealized models, appears to have bothered most Einstein and Schrödinger, as well as many others throughout the history of the debate concerning quantum mechanics. Indeed this aspect of quantum mechanics may be seen as defining this debate most essentially. Einstein and Schrödinger would prefer all physics to remain governed by the program of an at least approximately and, hopefully, asymptotically realist conceptual idealization of nature, however complex the nature of this idealization might be, and both expected it to be complex. As discussed earlier, they were well aware of the approximate, idealized, and imperfect, and as such subject to adjustment and replacement, character of our physical concepts and models. Einstein, who famously did not think that God, subtle though he is, would deceive us (not too much in any event!), also did not believe that God would want to conceal that much from us, in essence just about everything that counts. At stake (these stakes are high) is what our models offer us: a certain continuity with nature, at one end, and with our thinking or argumentation concerning (idealized) objects and processes, at the other, or, as in quantum theory, a radical break and discontinuity with both. In sum, the question is whether physics can offer us descriptive or at least explanatory arguments concerning idealization in conjunction with mathematical theories and models, or only use mathematics for the purposes of correct predictions in the absence of any connections between this mathematics and the objects and processes considered, even if the latter are idealized rather than actual. As discussed earlier, it is still possible to take a view of classical physics and, especially, relativity as ultimately offering only predictive rather than descriptive models of physical reality. Even in this case, however, the nature of these models is essentially different since those of quantum mechanics (at least as complementarity) do not offer mathematical models representing, moreover causally, objects and motion, even if the latter are seen as ideal ("fictional") rather than real or approximating reality.

These considerations also bear on the question of the relationships, similarities and differences, between Galileo's and Bohr's philosophies of physics or, again, of nature itself. Both are similar in that neither Galileo's nor Bohr's argumentation *offers* a philosophy of nature, a view concerning the ultimate nature of nature. While, however, Galileo's physics may not offer any claims concerning how nature actually works, and hence, as concerns nature, may be primarily predictive rather than descriptive (certain parts of Galileo's physics and astronomy are more descriptive), Galileo's mathematical-physical models themselves, enabling these predictions, are classical in character, and specifically causal. They are, it is worth noting, also, and correlatively, geometrical and (geometrically) diagrammatic in nature. They are, at the very, least phenomenalizations or "fictions" in the character that came to classical physics as (causal) motions of bodies. As such, they *suggest* at least a possibility of an idealized representation of natural processes, such as the motion of planets moving around the sun, which representations and corresponding models were developed with and following Newton, and were anticipated earlier, as in Copernicus and Kepler, and in some respects even earlier, beginning indeed at least with Aristotle's physics. Galileo's approach could be and was converted to a more Newtonian model of an idealized representation of natural bodies and motions, such as that of planets moving around the sun, which we are both (in classical physics) lucky and (in quantum physics) unlucky to have as our primary model.

In other words, it is the character of the models and hence, it follows, of the arguments concerning them, including those concerning the possibility or impossibility of certain types of models, that is the main issue here. There are literally no such physical or mathematically geometrical, or visualizable or intuitable (*anschaulich*) models in quantum mechanics, at least, as complementarity, no geometry in this broad sense, only algebra, enabling correct predictions, again, in general statistical in nature. Nonclassically, there isn't representation in quantum mechanics, or, when there is, it refers only to that which cannot be represented or even conceived of by any means. John von Neumann at some point said that "the sciences do not try to explain," but rather they "mainly make [mathematical] models," that is "mathematical construct[s] [that], with the addition of certain verbal interpretations, [describe] the observed phenomena" (von Neumann 1961-1963, v. 6, p. 491). This is precisely what is, as I argue here, no longer offered or even possible in quantum mechanics, at least as complementarity, even though, I also argue, mathematics itself, mathematics without physical models, even if ideal or "fictional" ones (i.e. again, without any correspondence between them and reality), remains as important as ever. Establishing this new type of relationships between mathematics and physics was one of Heisenberg's greatest and most revolutionary contributions. For, as just noted and as explained throughout this study, the mathematical formalism of the theory, which, we recall, was given its rigorous Hilbert-space shape by von Neumann, no longer describes either quantum objects or processes themselves. The latter indeed no longer form phenomena considered in quantum mechanics or can even form phenomena as anything that can possibly be observed, described, or even conceived by any means, mathematical or verbal included. Nor does this formalism describe what is actually observed, what constitutes the phenomena in question in quantum mechanics, which are the products of the (quantum) interactions between quantum objects and measuring instruments and as such are described by means of classical physics, accompanied by a certain verbal interpretation. The latter is an important qualification, rightly added by von Neumann, and it is rather Bohrian in spirit, and perhaps follows Bohr. This qualification remains valid and, as Bohr makes clear, even more essential in quantum mechanics, even in and in part by virtue of the absence of mathematical models describing the emergences of these phenomena. Classical physics, however, cannot predict such phenomena or of course explain them or describe their emergence (although explanation and description are closely connected in classical physics).

In sum, both views, Galileo's and Bohr's, are *consistent* with nonclassical epistemology, but only Bohr's rigorously *conforms* to it, and, at the very least, suggests that, ultimately, nature *may not be representable or conceivable* and, hence, that it *may require* nonclassical theories from us. Accordingly, although the ultimate (which is, again, to say, next) jury is still out, a Newtonian conversion of the type just mentioned may not be possible in the case of quantum theory or even nature itself.

That this may not be possible makes an immense difference, including as concerns possible claims, claims *in principle*, regarding nature. In particular, while the principles of science outlined above define *both* classical and quantum physics, in order to rigorously maintain them in the case of quantum physics, quantum phenomena and quantum theory alike, one must, according to Bohr, accept the nonclassical epistemology of both. If one does so, however, one must also abandon, at the level of quantum objects and processes, certain other (epistemological and ontological) principles, applicable (alongside the basic principles of science as just outlined) in classical physics. The (epistemological) nonclassicality of quantum physics becomes, rigorously, an essential condition of its disciplinary character as physics, of the continuity of the practice of physics as a mathematical science of nature. Thereby it also establishes a continuity between quantum and classical physics, which would be broken otherwise. It does so, however, by radically departing from classical epistemology and, hence, as Bohr says in his reply to EPR, from "the customary viewpoint of natural philosophy, ... an essential inadequacy of [which] for a rational account of physical phenomena with which we are concerned in quantum mechanics is disclosed by the apparent contradiction [found by EPR]" (*PWNB* 4, pp. 74-75).[75] In sum, the break from classical physics occurs at the level of epistemology, not at that of the character and the practice of physics as a mathematical science or, since both define each other in science, as a mathematical-experimental science, as both Einstein and Heidegger argue. Einstein, of course, had a very different way of linking mathematics and experiment in mind. Heidegger's views are more complex and he, actually, assesses quantum mechanics and especially Heisenberg's work in positive terms. As I have stressed throughout, classical epistemology is not abandoned either. Along with classical science it continues to function within its proper limits and is often part of nonclassical theories as well, which often depend on it. Under the nonclassical conditions, however, the laws of quantum physics are the laws of nature only in the sense of corresponding to the "regularities" that nature allows to our interaction with it, specifically by means of experimental technology. The term is used by Bohr in speaking of "the new types of regularities," the ones that we encounter as effects of the interaction between the ultimate quantum constituents of nature and our measuring technology, and which cannot be accounted for by classical physics (*PWNB* 4, p. 81). There is nothing that quantum theory, or in view of its laws, conceivably any theory, can say about these constituents as such. In the nonclassical interpretation given to it by Bohr's complementarity, quantum mechanics represents the interactions between what is representable by classical physical and epistemological means, specifically in measuring instruments, described by the laws of classical physics (which allow for a realist and causal interpretation), and what is unrepresentable by any means available to us, or is even ultimately inconceivable. On the present definition of nonclassical theory, which Bohr's complementarity paradigmatically establishes, whatever can, in principle, be represented is considered epistemologically classical; and the quantum-mechanical situation of complementarity just defined is generalizable to nonclassical theories elsewhere and, as explained earlier, indeed defines nonclassicality itself. This is a

---

[75] One can, of course, practice (in the technical or mathematical-experimental sense) quantum physics while subscribing to the classical philosophy of nature or of physics, including as concerns quantum theory itself. The question is whether one can explain classically the nature of quantum phenomena or quantum theory, or (a more complex issue) the nature of this practice or the practice of quantum theory in general.

radical, revolutionary change in our view of nature and science, or of knowledge in general.

Contrary to common claims and some appearances, however, in Bohr's case (or that of Heisenberg or several others who may be invoked here, Pauli, for example), one encounters what might be called the extreme disciplinary conservatism or, extending the notion, the extreme conservatism of theoretical thinking. I mean by this an extreme reluctance to bring in a radical change, which is finally done only at points and in regions where there is really no choice, in the sense that their discipline (in either sense) in fact requires it. A departure from a given preceding configuration of thought is enacted, first, after exhausting the possibilities it offers for a new configuration, which may in fact arise in part from within the old one, and, secondly, under the extreme pressure of maintaining and even conserving significant disciplinary or even epistemological-philosophical aspects of the old configuration.

I borrow the phrase "extreme conservatism" from Sylvan S. Schweber, although I give it a broader and somewhat different meaning. In his *QED and the Men Who Made It* (Schweber 1994), Schweber argues that in the case of quantum electrodynamics, it was the persistence in keeping the existing framework, with incremental modifications, rather than attempts at radically transforming it, that paid off. In the case of QED it was, ironically, Dirac, its founder, who gave up on his creation and believed that yet another radical transformation, similar to that of the original quantum mechanics in relation to classical physics, would be necessary. Schweber speaks of the "extreme conservatism" of the figures mentioned in his title in this context and in this sense. From the present perspective, the extreme conservatism may apply even when a radical transformation is ultimately at stake and is deliberately attempted, in contrast, for example, to the case of Planck's discovery of the quantum. There, its momentous revolutionary nature *may be* seen as having become apparent more in retrospect and was even helped by the thought of others, in particular Einstein. Well, it may be seen in this way. But should it be? It is difficult to be certain. Planck, conservative by nature, might have been a "reluctant revolutionary," as some argue (the phrase comes courtesy of A. Pais). But he made a revolutionary move nevertheless, upon exhausting all his classical-physical resources. Indeed, he made some revolutionary moves in his earlier work in thermodynamics. Such situations always have great complexities to them, and it would be difficult to argue that extreme conservatism is necessary in all conditions or at all points, even in science, although virtually all the founders of quantum mechanics appeared to conform to this view at the time of its emergence. We can never be certain what will ultimately pay off. As was indicated in the preceding chapter, in some respects the creation of modern QED was quite radical as well, in particular in employing rather unorthodox, and indeed mathematically strictly forbidden, techniques in the renormalization procedure, the centerpiece of quantum field theory ever since. So the creators or (it was founded by Dirac and several others earlier) perhaps "saviors" of modern QED, too, were both extreme conservatives and extreme radicals, just as were the founders of quantum mechanics, certainly both Bohr and Heisenberg, or differently, Einstein.

Thus, while the situation has more general implications, Bohr derives his nonclassical epistemology from his analysis of the character of quantum phenomena themselves and of the mathematical-experimental structure of quantum mechanics, under the type of pressure invoked, and this is in part why he speaks of the epistemological lesson of quantum mechanics. He had no choice but to move in this direction, just as was the case earlier in his 1913 theory of the hydrogen atom or in Heisenberg's discovery of quantum mechanics, both of which made earlier steps on the long road to nonclassical epistemology. Einstein was similarly compelled to his radical moves in the cases of both special and general relativity, although along more classical lines, or so it appeared to him at the time. One is not so sure anymore. According to Heisenberg himself (writing in 1934), in the case of new physics (relativity and quantum mechanics): "Modern theories did not arise from revolutionary ideas which have been, so to speak, introduced in the exact sciences from without. On the contrary they have forced their way into research which was attempting consistently to carry out the programme of classical physics—they arise out of its very nature. It is for this reason that the beginning of modern [twentieth-century] physics cannot be compared with the great upheavals of previous periods like the achievements of Copernicus" (Heisenberg 1979, p. 13). The point concerning the time of Copernicus may require further qualification, since the rise of modern science may contain more continuity with the preceding history than Heisenberg's strong claim may suggest. Heisenberg is, however, correct as concerns the emergence of quantum mechanics, including his own discovery of it, even though and because it was so revolutionary, physically and, with some help from Bohr, philosophically.

6. CONCLUSION: CHAOSMIC ORDERS

In Bohr and Heisenberg alike, physics and philosophy are brought together by linking a form of radical (in Bohr's case, nonclassical) epistemology to the invention of new concepts, the primary business of theoretical thought in physics and philosophy alike, through which it confronts chaos and creates order in this confrontation. As an experimental-mathematical science of nature, as defined itself from Galileo on, physics needs mathematics, and mathematics sometimes needs physics to move forward, but both may need philosophy more than they sometimes like to think. Philosophy, too, needs mathematics and physics more than it sometimes thinks. Most major philosophical figures, beginning with Plato (a more mathematically oriented thinker) and Aristotle (a more physically oriented thinker) on, knew better, and quite a few were of course also mathematicians or physicists or both, with Descartes and Leibniz as primary examples, and many more were trained in mathematics or physics. Indeed, philosophy has always appeared to have an ambition to become like mathematics or even a form of mathematics, and sometimes physics. This is, admittedly, more rare, although Aristotle, again, is a primary example, or in some respects Nietzsche. As noted earlier, however, Nietzsche, too, urged philosophy to become mathematics to the degree possible, albeit for different, more epistemologically nonclassical, reasons (Nietzsche 1974, p. 215). As Hermann Weyl says in closing his introduction to his great book, *Space Time Matter*, also thinking of his own task in the book:

> All beginnings are obscure. Inasmuch as the mathematician operates with his conceptions along strict and more formal lines, he, above all, must be reminded from time to time that the origins of things lie in greater depths than those to which his methods enable him to descend. Beyond knowledge gained from the individual sciences, there remains the task of *comprehending*. In spite of the fact that the views of philosophy sway from one system to another, we cannot dispense with it unless we are to convert knowledge into a meaningless chaos. (Weyl 1921, p. 10)

The obscurity of all beginnings is one of those great philosophical perceptions that are characteristic of Weyl's writings, in some of which Weyl goes much further in his appeal or (given a common skepticism on the part of many scientists) defense of philosophy and its significance for mathematics and science, or, again, reciprocally, their significance for philosophy. Weyl's thinking, especially in his major works, *Time Space Matter* included, are steeped in the phenomenology of Franz Brentano, Edmund Husserl, and Henry Bergson, as well as in earlier ideas of such figures as Johann Gottlieb Fichte, a philosophical tradition, extending from Kant, which I stressed in this chapter. Well aware of scientists' skeptical attitude just mentioned, Weyl does not fail to offer perceptive, if ironic, observations and prudent disclaimers. Thus, as he notes in *The Continuum*: "We cannot set out here in search of a definitive elucidation of what is to be a state of affairs, a judgment, an object, or a property [the concepts that are usually not critically explored or even noticed as concepts by mathematicians and scientists, who nonetheless use them all the time]. This task leads into metaphysical depths. And concerning it one must consult men, such as Fichte, whose names may not be mentioned among mathematicians without eliciting an indulgent smile" (Weyl 1918, p. 7).

On the other hand, insofar as we can know or can conceive of it, chaos is not only an enemy but also a friend of thought, as Deleuze and Guattari argue in *What is Philosophy?* (Deleuze and Guattari 1993). Chaos, they further argue, is also thought's greatest ally in its struggle against opinion, *doxa*, always an enemy only, as much in mathematics and science, or philosophy, as elsewhere in life, "like a sort of 'umbrella' that protects us from chaos." This is, for example, how we often think or use already established theories, for example, in physics, although, as I argued before, a certain, even extreme, conservatism (not the same as opinion, however) is not so bad sometimes, including on our way to new theories and, thus, new confrontations with chaos. "But," Deleuze and Guattari say, "art, science, and philosophy require more: they cast planes over chaos [...] The struggle with chaos is only the instrument in a more profound struggle against opinion, for the misfortune of people comes from opinion" (p. 202). They add: "And what would *thinking* be if it did not confront chaos?" (p. 208).

Deleuze and Guattari approach chaos by means of a particular and, in philosophy, rarely, if ever, used concept. According to Deleuze and Guattari: "Chaos is defined not so much by its disorder as by the infinite speed with which every form taking shape in [the phenomenal field of thought] vanishes. It is a void that is not a nothingness but a *virtual*, containing all possible *particles* and drawing out all possible forms, which spring up only to disappear immediately, without consistency or reference, without consequence. Chaos is an infinite speed of birth and disappearance" (Deleuze and Guattari 1994, p. 118; emphasis on "particles" added). Although unusual in philosophy, this type of idea of chaos, or at least of *virtuality*, transpires in and is in part derived by Deleuze and Guattari from quantum field theory (Deleuze and Guattari 1994, p. 225, n.1). It relates to the concept, discussed in Chapter 5, of virtual creation and annihilation of particles from the false vacuum, a sea of energy, thus suggesting the image of chaos invoked by Deleuze and Guattari.

These two concepts are not identical. For one thing, the speed of all processes in question in quantum field theory is finite and limited by the speed of light in a vacuum. Deleuze and Guattari's appeal to infinite speed relates (one might say, metaphorically, if one keeps in mind that once it is transferred into Deleuze and Guattari's philosophy, 'speed of thought' becomes a philosophical concept for them) to the speed of thought or of the emergence of images in the field of thought. The term chaos itself is not used in quantum field theory, although some chaos may be detected beneath the use of Feynman diagrams, which, in Deleuze and Guattari's view, would "slow down" the actual processes of virtual creation and annihilation of particles in question in quantum field theory. At any point of the processes "represented" by a given Feynman diagram yet another virtual process may occur and hence another diagram may be inserted into it, thus leading to an interminably expandable topological structure. Feynman diagrams are, however, just diagrams, pictures that help us to heuristically visualize the situation, or, again, to *slow down* the phenomenological chaos of the situation, to hold in mind the forms thus created, for the purposes of helping calculations. So is the "picture" (conception) of the virtual particle formation. What actually happens at the level of such processes themselves we might no more know or even conceive of, let alone visualize, than we can in quantum mechanics, and, as discussed in Chapter 5, quantum field theory may imply even greater epistemological and conceptual complexities than those found in quantum mechanics. It may, however, also eventually lead, via string and brane theories, for example, to a more classical and specifically more visualizable picture of an underlying order. For the moment, however, quantum field theory appears, at the very least, to be more conducive to a more quantum-mechanical and, hence, from the present perspective, nonclassical view of the ultimate constitution of matter.

This view of quantum field theory allows one to link the concept of *chaos as the virtual* to the idea of *chaos as the incomprehensible*, which can be traced to the Ancient Greek idea of chaos as *areton* or *alogon*—that which is beyond all comprehension. Yet another concept of chaos, *chaos as disorder*, defined by the role of chance in its workings or in its effects, may and even must be invoked here as well. As Deleuze and Guattari's formulation indicates, the concept of chaos as disorder is not entirely put aside by them: while "chaos" may be "defined *not so much* by its disorder," it may partially be defined by disorder or, at least, by chance.

I would not claim that these three concepts of chaos are the only ones possible or even available. I would argue, however, that they—all three of these concepts (rather than, as Deleuze and Guattari suggest, only the first one)—are operative in our understanding of the nature of thought and its struggle against opinion, including in science and specifically in quantum physics, from quantum mechanics to quantum field theory and beyond. It is remarkable that, as follows from the analysis given in this book, in confronting the initial chaos of quantum phenomena and, with it, also the *doxa*, opinion, of classical physics, which could not handle this chaos, quantum theory brought order into this chaos in part by making chaos itself, in all three senses of it, part of this order. I hasten to qualify, before explaining why such is the case, that one can speak of the *thought* of classical physics as *doxa* only in this context. In general, just as that of quantum physics, the thought of classical physics is defined by its many confrontations with chaos of phenomena or ideas. Besides, as I stressed throughout this study, we still depend on classical physics, including in quantum theory. For the moment, first, the fact that in quantum mechanics and quantum field theory alike, at least as complementarity, we might no more know or even conceive of, let alone visualize the ultimate objects and processes in question, implies the essential presence of chaos as the incomprehensible in quantum field theory. Since, in addition, all our knowledge concerning the ultimate constitution of nature is only predictive and, moreover, only statistical, chaos as chance and disorder is added to the picture as well. Indeed, as we have seen, the very character of chance in quantum physics as irreducible to any underlying or hidden necessity arises by virtue of the ultimately unknowable and even inconceivable nature of quantum objects and processes, as these are seen or idealized by quantum theory. In short it arises by virtue of the role of chaos as the incomprehensible in the conceptual architecture of quantum theory.

This is how chaos (of both quantum nature and of the mind confronting it), chaos as the incomprehensible and chaos as chance and disorder, was approached by quantum mechanics, beginning at least with Bohr's 1913 theory of the atom, and it helped quantum mechanics against the *doxa* of classical physics. Thus *chaos*, chaos as the incomprehensible and chaos as chance or disorder, indeed becomes part of a new *order* of quantum physics, that is, of the order it finds in nature, or part of its chaosmos, to use James Joyce's coinage, favored by Deleuze and Guattari as well. In quantum field theory, these two concepts of chaos are retained, but they are, as we have seen, not sufficient to deal with the chaos quantum field theory has to confront and to build its physical and mathematical architecture. To accomplish this task, quantum field theory is compelled to engage with chaos as the virtual, along with chaos as the incomprehensible and chaos as randomness and chance. As discussed in Chapter 5, this extension of the conceptual architecture of quantum theory, itself a major conceptual invention, may (the epistemology of quantum field theory is, as I noted, as yet a barely explored subject) entail yet further epistemological complexities, as against those of quantum mechanics.

The reasons for this situation and for the necessity of bringing chaos (in these three senses, and conceivably others) into the order of quantum theory appears to be the irreducible role of measuring instruments in forming the data, the phenomena (also in Bohr's sense) in question in the theory. This line of thought guides Bohr's argument for complementarity from the Como lecture on. To cite, for the last time in this study, Bohr's arguably most refined formulation, this situation arises by virtue of "the *impossibility of any sharp separation between the behavior of atomic objects and the interaction with the measuring instruments which serve to define the conditions under which the phenomena appear*" (Bohr 1949, *PWNB* 2, pp. 39-40). As discussed in Chapter 5, it appears that further complexities of quantum field theory, including the emergence of the concept of the virtual particle formations and hence the role of chaos of the virtual, are, too, due to the difficulties we have in handling, physically and, hence, also mathematically, and epistemologically, these interactions in the (high-energy) relativistic regimes. These difficulties appear insofar as we have, perhaps unavoidably, to suspend in our analysis the quantum constitution of measuring instruments, even though this constitution is essential for our quantum experiments, since otherwise no interaction between quantum objects and measuring instruments would be possible. This interaction is obviously essential in quantum mechanics, which, however, appears to allow us to disregard the quantum aspects of measuring instruments, while this may be no longer possible in the case of the experiments in question in quantum field theory.

As it has often done before, including in classical physics, physics had to plunge into the chaos to be able to create the order of quantum mechanics. As Bohr noted: "The history of physical science thus demonstrates how the exploration of ever wider fields of experience, in revealing unsuspected limitations of accustomed ideas [*doxa*], indicates new ways of restoring logical order" (*PWNB* 2, p. 74). Bohr also notes that this situation and this attitude may be generalized beyond physics. He sees it as "the endeavour to achieve a harmonious comprehension of ever wider aspects of our [human] situation, recognizing that no experience is definable without a logical frame and that any apparent disharmony can be removed only by an appropriate widening of the conceptual framework" (*PWNB* 2, p. 81). In the case of quantum theory, however, this new order of theory retains chaos as part of the understanding of nature at the quantum level of its constitution that the theory offers. For, as Bohr also says at the same juncture, "in the complementary description of quantum physics, we have to do with a further self-consistent generalization which permits the inclusion of regularities decisive for the account of fundamental properties of matter, but which transcends the scope of deterministic [or realist] description" (*PWNB* 2, p. 74). Accordingly, this self-consistent generalization makes chaos as chance and chaos as the incomprehensible, and in the case of quantum field theory, chaos as the virtual, part of what theory tells us about nature.

In other words, the thought of quantum theory, at least as the nonclassical thought of complementarity, confronts chaos by making the idea of chaos itself part of the order of the theory, part of its thinking and knowledge. This is of course the situation that defines all nonclassical thought, wherever it is found, although in quantum theory, just as in classical physics, it is mathematics that allows thought to handle the confrontation with chaos. This may in turn appear as a miracle, although it may also be due to the fact that this is how, throughout the mathematization of whatever we can mathematize, we setup modern physics. In quantum theory, however, this is still epistemologically mysterious, since, unlike in classical physics, the mathematical formalism enables correct (statistical) predictions with no mathematical description, however idealized, of the ultimate processes responsible for the rise of the phenomena in question. It is worth reiterating, however, that, while nonclassicality forms a consistent concept (in this case of a possible theoretical framework) and while its justification is, at least in the present view, *theoretical* rather than merely or only practical (in Kant's sense explained earlier), it is *interpretive*. That is, this justification is provided by the fact that, as part of complementarity, nonclassicality enables one to offer a consistent interpretation of quantum mechanics as a complete and local theory. Nature itself, at the ultimate level of its constitution may or may not be classical, and, at least at certain future stages of physics, a return to classicality is not inconceivable. I mean a return across the board, since classicality suffices in much other physics, although for the moment it does not appear (some, again, argue otherwise) that does in quantum physics, from quantum mechanics on. On the other hand, a move in the opposite direction, away from classical epistemology but beyond nonclassical epistemology as here defined, is not inconceivable either. The justifications or significance of nonclassicality beyond physics, for example, in mathematics, is a separate issue that cannot be considered here.[76] Suffice it to restate for the moment that it forms a consistent general philosophical concept, with a possible field of applications, and that quantum mechanics as complementarity demonstrates this consistency and this possibility of application.

---

[76] On this point, I permit myself to refer to (Plotnitsky 1994 and Plotnitsky 2002).

It is thus crucial that, while it contains an essentially Hegelian dimension insofar as it involves the invention of new concepts and conceptual frameworks, complementarity or any nonclassical theory extends and radicalized the Kantian epistemological dimension. It does so by retaining the unknowable and, beyond or, again, radicalizing Kant, the inconceivable, chaos as the incomprehensible, as part of the overall framework the theory offers. One might also argue that it is this radicalization of Kant into full-fledged nonclassicality that disallows the situation to be subsumed by Hegel's conceptuality, ultimately classical in nature. Nor, and in part correlatively, can the conceptual-epistemological situation of quantum mechanics be subsumed by a Hegelian conceptual-historical dialectic (it is, by definition, interactively both, conceptual and historical), although it may contain elements of this dialectic within certain limits, for example, insofar as various aspects of classical physics or classical epistemology are retained in quantum mechanics and complementarity. Overall, however, it is not a question of a contradiction or a set of contradictions between classical and quantum physics, or classical and nonclassical epistemology, or of a synthesis resolving these contradictions, which process defines Hegelian dialectic. A more complex set of relationships obtains here, and some of the contradictions or incompatibilities between the regimes involved are not and might never be resolved. On the other hand, as I noted earlier, it may depend on how one reads Hegel, or indeed Bohr, who sometimes invokes dialectic or synthesis, but, I would argue, only within certain limits, and without a rigorous philosophical elaboration of these concepts of the kind one finds in Hegel (*PWNB* 2, p. 63).

In any event, in the present reading, the nonclassical nature of complementarity moves it beyond Kant and Hegel alike, although it might not quite move it beyond the historical situation defined by the great confrontation between Kant and Hegel, the situation that is still our own, and within which the confrontation between Bohr and Einstein takes place. Thought remains a confrontation with chaos, but it needs to make chaos part of the theories it creates to give order to our encounter with nature, and, it appears, with mind as well, although in this materialist view, mind, the product of certain material, including physical, processes in our brains, the product of our bodies, is still nature.

It is possible to see the nonclassical view of thought and knowledge, at least in the case of quantum physics, as a new *doxa*, a *doxa* of the orthodox (Copenhagen) interpretation. It is also possible, however, with Bohr and Heisenberg, to see it as part, but only a part, of the experimentally justified axiomatic of quantum theory (both quantum mechanics and quantum field theory). Far from preventing the development of quantum theory as, in Bohr's words, "regards experimental evidence and the mathematical tools appropriate to its comprehension," the nonclassical epistemology of this axiomatic enables this development and, thus, the work of thought to continue, in its confrontation with chaos (*PWNB* 3, p. 6). Similarly to the great ocean, the chaos of the great ocean, in the ancient Greek conception of the world, this chaos surrounds the small archipelago of our thought and knowledge, and two sub-archipelagoes, which share islands between each other, and some of their islands with mathematics, or other sciences or art, and which we call physics and philosophy.

References

ABBREVIATIONS

PWNB: Bohr, N., *The Philosophical Writings of Niels Bohr*, 3 vols., Woodbridge, CT: Ox Bow Press, 1987; Bohr, N., *The Philosophical Writings of Niels Bohr, Volume 4: Causality and Complementarity, Supplementary Papers*, ed. Faye, J. and Folse, H. J, Woodbridge, CT: Ox Bow Press, 1998.

QTM: Wheeler, J. A. and Zurek, W. H. *Quantum Theory and Measurement*, Princeton, NJ: Princeton University Press, 1983.

Bell, J. S.

1987 *Speakable and Unspeakable in Quantum Mechanics*, Cambridge: Cambridge University Press, 1987.

Beller, M.

1999 *Quantum Dialogue: The Making of a Revolution*, Chicago: University of Chicago Press, 1999.

Bertlmann, R. A. and Zeilinger, A., eds.

2002 *Quantum (Un)speakables: From Bell to Quantum Information*, Berlin: Springer, 2002.

Bohm, D.

1995 *Wholeness and Implicate Order*, London: Routledge, 1995.

Bohr, A., Mottelson, B. R., and Ulfbeck, O.

2004 "The Principles Underlying Quantum Mechanics," *Foundations of Physics* 34 (3) (2004), 405-517.

Bohr, N.

1935 "Quantum Mechanics and Physical Reality," in Wheeler, J. A. and Zurek, W. H. *Quantum Theory and Measurement*, Princeton, NJ: Princeton University Press, 1983.

1972-1996 *Niels Bohr: Collected Works*, 10 vols., Amsterdam: Elsevier, 1972-1996.

1987 *The Philosophical Writings of Niels Bohr*, 3 vols., Woodbridge, CT: Ox Bow Woodbridge Press, 1987.

1998 *Philosophical Writings of Niels Bohr, Volume 4: Causality and Complementarity, Supplementary Papers*, ed. Faye, J. and Folse, H. J, Woodbridge, CT: Ox Bow Press, 1998.

Born, M.

2005 *The Einstein-Born Letters*, trans. Born, I., New York: Walker, 2005.

Born, M., Heisenberg, W., Jordan, P.

1926 *Zur Quntenmechanik*. II, in Born, M., Heisenberg, W., and Jordan, P., *Zur Begründung der Matrizenmechanik (Dokumente der Naturwissenschafften —Abteilung Physik, Vol. 2*, Hermann, A. ed.), Stuttgart: E. Battenberg Verlag, 1962.

Brown, H. R. and Pooley, O.

2001 "The Origin of Spacetime Metric: Bell's 'Loretzian Pedagogy' and its Significance in General Relativity," in Callender, C. and Huggett, N., *Physics Meets Philosophy at the Planck Scale: Contemporary Theories of Quantum Gravity*, Cambridge: Cambridge University Press, 2001, 256-272.

Butterfield, J. and Isham C.

2001 "Spacetime and the Philosophical Challenge of Quantum Gravity," in Callender, C. and Huggett, N., *Physics Meets Philosophy at the Planck Scale: Contemporary Theories of Quantum Gravity*, Cambridge: Cambridge University Press, 2001, 33-89.

Cartier, P.

2001 "A Mad Day's Work: From Grothendieck to Connes and Kontsevich. The Evolution of Concepts of Space and Symmetry," *Bulletin (New Series) of the American Mathematical Society* 38 (4) (2001), 389-408.

Connes, A.

1994 *Noncommutative Geometry*, trans. Berberian, S. K., ed. Rieffel, M. A., San Diego, CA: Academic Press, 1994.

Cushing, J. T. and McMullin, E., ed.

1989 *Philosophical Consequences of Quantum Theory: Reflections on Bell's Theorem*, Notre Dame, IN: Notre Dame University Press, 1989.

D' Espagnat, B.

1989 *Conceptual Foundations of Quantum Mechanics*, Redwood City, CA: Addison-Wesley, 1989.

Dirac, P. A. M.

1933 Letter to Niels Bohr, August 10, 1933, copy in Niels Bohr Library, Center for the History of Physics, American Institute of Physics, College Park, Maryland.

1962 T. Kuhn, interview with Dirac, April 1, 1962, Niels Bohr Library, Center for the History of Physics, American Institute of Physics, College Park, Maryland.

1995 *The Principles of Quantum Mechanics*, Oxford: Clarendon, 1995.

Dyson, F. J.

1949 "The *S*-Matrix in Quantum Electrodynamics," *Physics Review* 75 (1949), 1736-1755.

2005 "Hans Bethe and Quantum Electrodynamics," *Physics Today* 58 (10) (2005), 48-50.

Einstein, A., Podolsky, B., and Rosen, N.

1935 "Can Quantum-Mechanical Description of Physical Reality be Considered Complete?", in Wheeler, J. A. and Zurek, W. H. *Quantum Theory and Measurement*, Princeton, NJ: Princeton University Press, 1983, 138-141.

Ellis, J. and Amati, D., eds.

2000 *Quantum Reflections*, Cambridge: Cambridge University Press, 2000.

Faye, J.

1991 *Niels Bohr: His Heritage and Legacy. An Anti-Realist View of Quantum Mechanics*, Dordrecht: Kluwer, 1991.

Faye J. and Folse, H. J., eds.

1998 *The Philosophical Writings of Niels Bohr, Volume 4: Causality and Complementarity, Supplementary Papers*, Woodbridge, CT: Ox Bow Press, 1998.

Fine, A.

1989 "Do Correlations Need to be Explained?", in Cushing, J. T. and McMullin, E., eds., *Philosophical Consequences of Quantum Theory: Reflections on Bell's Theorem*, Notre Dame, IN: Notre Dame University Press, 1989, 174-94.

Feynman, R.

1988 *QED: A Strange Theory of Light and Matter*, Princeton: Princeton University Press, 1988.

Folse, H, J.

1985 *The Philosophy of Niels Bohr: The Framework of Complementarity*. Amsterdam, North Holland, 1985.

1987 "Niels Bohr's Concept of Reality," in Pekka Lahti and Peter Mittestaedt, eds., *Symposium on the Foundations of Modern Physics 1987: The Copenhagen Interpretation 60 Years after the Como Lecture*, Singapore: World Scientific, 1987, 161-80.

2002 "Bohr's Conception of the Quantum-Mechanical State of a System and Its Role in the Framework of Complementarity," in Khrennikov, A., ed. *Quantum Theory: Reconsiderations of Foundations 2001*, Växjö: Växjö University Press, 2002, 83-98.

Friedman, J. R., Patel, V., Chen, W., Toltygo, S. K., and Lukens, J. E.

2002 "Quantum Superposition of Distinct Macroscopic States," *Nature* 406 (2000), 43-46.

Fuchs, C. A.

2001 "Quantum Foundation in the Light of Quantum Information," in Gonis, T. and Turchi, P.E.A., ed., *Decoherence and Its Implications in Quantum Computation and Information Transfer*, Amsterdam: IOS Press, 2001, 38-82

2003 "Quantum Mechanics as Quantum Information, Mostly," *Journal of Modern Optics* 50 (2003), 987-1003.

Fuchs, C. A. and Peres, A.

2000 "Quantum Theory Needs No 'Interpretation,'" *Physics Today*, 53 (3) (2000), 70.

Garg, A.

2001 "Prospects for Macroscopic Quantum Coherence," in Gonis, T. and Turchi, P.E.A. Turchi, ed., *Decoherence and its Implications in Quantum Computation and Information Transfer*, Amsterdam: IOS Press, 2001, 256-283.

Greene, B.

1999 *The Elegant Universe: Superstrings, Hidden Dimensions, and the Quest for the Ultimate Theory*, New York: W. W. Norton, 2003.

Gottfried, K.

2000 "Does Quantum Mechanics Carry the Seeds of Its Own Destruction," in Ellis, J. and Amati, D. eds., *Quantum Reflections*, Cambridge: Cambridge University Press, 2000, 165-185.

Grattan-Guinness, I.

1998 *The Norton History of Mathematical Sciences: The Rainbow of Mathematics*, New York: W. W. Norton, 1998.

Griffiths, R. B.

2003 *Consistent Quantum Theory*, Cambridge: Cambridge University Press, 2003.

Hacking, I.

1984 *Emergence of Probability*, Cambridge: Cambridge University Press, 1984.

Haroche, S.

2001 "Entanglement and Decoherence in Cavity Quantum Electrodynamics Experiments," in Gonis, T. and Turchi, P. E.A., eds., *Decoherence and Its Implications in Quantum Computation and Information Transfer*, Amsterdam: IOS Press, 2001, 211-223.

Haroche, S., Brune, M., and Raimond, J. M.

1997 "Experiments with Single Atoms in Cavity: Entanglement, Schrödinger's Cats, and Decoherence," *Philosophical Transactions of the Royal Society of London*, 355 (1997), 2367-2380.

Hawking, S. and Penrose, R.

1996 *The Nature of Space and Time*, Princeton, NJ: Princeton University Press, 1996.

Hegel, G. W. F.

1977 *Phenomenology of Spirit*, trans. Miller, A. W., Oxford: Oxford University Press, 1977.

1990 *Hegel's Science of Logic*, trans. Miller, A.W., Atlantic Highlands, NJ: Humanities Press International, 1990.

Heisenberg, W.

1925 "Quantum-Theoretical Re-Interpretation of Kinematical and Mechanical Relations," in Van der Waerden, B. L., ed. *Sources of Quantum Mechanics* Toronto: Dover, 1968, 261-77.

1927 "The Physical Content of Quantum Kinematics and Mechanics," in Wheeler, J. A. and Zurek, W. H. eds., *Quantum Theory and Measurement*, Princeton, NJ: Princeton University Press, 1983, 62-86.

1930 *The Physical Principles of the Quantum Theory*, trans. Eckhart, K. and Hoyt, F. C., New York: Dover, 1930, rpt. 1949.

1967 "Quantum Theory and its Interpretation," in Stephan S. Rozental, *Niels Bohr: His Life and Work as Seen by his Friends and Colleagues*, Amsterdam: North-Holland, 1967.

1979 *Philosophical Problems of Quantum Physics*, Woodbridge, Conn.: Ox Bow Press, 1979.

1983 *Encounters with Einstein*, Princeton, NJ: Princeton University Press, 1983.

Heidegger, M.

1967 *What is a Thing?*, tr. W. B. Barton, Jr., and Vera Deutsch, South Bend, In.: Gateway, 1967.

Hoffmann, B.

1972 *Albert Einstein: Creator and Rebel*, London: Hart-Davis, 1972.

Holevo, A. S.

1973 "Information-Theoretical Aspects of Quantum Measurement," *Problems of Information Transmission* 9 (1973), 110-18.

Honner, J.

1987 *The Description of Nature: Niels Bohr and the Philosophy of Quantum Physics*, Oxford: Clarendon, 1987.

Iorio, A., Lambiase, G., and Vitiello, G.

2002 "Hopf's Algebras, Thermodynamics and Entanglement in Quantum Field Theory," quant-ph/020740, 8 July 2002.

Khrennikov, A. Yu.

2004 "Reconstruction of Quantum Theory on the Basis of the Formula of Total Probability," in Khrennikov, A. Yu., ed. *Foundations of Probability in Physics 3 (AIP Conference Proceedings, v. 750)*, Melville, NY: American Institute of Physics, 2004, 187-211.

Kuhn, T.

1962 *The Structure of Scientific Revolutions*, Princeton: Princeton University Press, 1962.

Leggett, A. J.

1988 "Experimental Approaches to the Quantum Measurement Paradox," *Foundations of Physics*, 18 (9) (1988), 939-952.

Mehra, J. and Rechenberg, H.

2001 *The Historical Development of Quantum Theory*, 6 vols., Berlin: Springer, 2001.

Mermin, N. D.

1990 *Boojums All the Way Through*, Cambridge: Cambridge University Press, 1990

1998a "What Is Quantum Mechanics Trying To Tell Us?", *American Journal of Physics* 66 (9) (1998), 753-767.

1998b "Nonlocal Character of Quantum Theory?", *American Journal of Physics* 66 (10) (1998), 920-924.

Mittelstaedt, P.

1987 "Language and Reality in Quantum Physics," in Lahti, P. and Mittelstaedt, P., ed. *Symposium on the Foundations of Modern Physics 1987*, Singapore: World Scientific, 1987, 229-250.

1994 "Kant and the Quantum Theory," Parrini, P., ed. *Kant and Contemporary Epistemology*, Dordrecht, The Netherlands: Kluwer, 1994.

1997 *The Interpretation of Quantum Mechanics and the Measuring Process*, Cambridge: Cambridge University Press, 1997.

Myatt, C. J., King B. E. , Turchette, Q. A., Sackett, C. A. , Kielpinski, D., Itano, W.M., Monroe, C., and Wineland, D. J.

2000 "Decoherence of Quantum Superpositions through Coupling to Engineered Reservoirs," *Nature* 403 (2000), 269-273.

Nietzsche, F.

1974 *The Gay Science*, trans. Walter Kaufmann, New York: Vintage, 1974.

Omnés, R.

1994 *The Interpretation of Quantum Mechanics*, Princeton, NJ: Princeton University Press, 1994.

1999 *Understanding Quantum Mechanics*, Princeton, NJ: Princeton University Press, 1999.

Pais, A.

1982 *Subtle is the Lord: The Science and the Life of Albert Einstein*, Oxford: Oxford University Press, 1982.

1986 *Inward Bound: Of Matter and Forces in the Physical World*, Oxford: Oxford University Press, 1986.

1991 *Niels Bohr's Times, In Physics, Philosophy, and Polity*, Oxford: Clarendon, 1991.

Pauli, W.

1979-1999 *Wissenschaflicher Briefwechsel,* Scientific Correspondence, Berlin: Springer, 1979-1999.

Penrose, R.

1994 *Shadows of the Mind: A Search for a Missing Science of Consciousness*, Oxford: Oxford University Press, 1994.

Peres, A.

1993 *Quantum Theory: Concepts and Methods*, Dordrecht: Kluwer, 1993.

1984 "What Is A State Vector?", *American Journal of Physics* 52 (1984), 644-650.

2002 "Karl Popper and the Copenhagen Interpretation," *Studies in History and Philosophy of Modern Physics* 33 (2002), 23-34.

2003 "Einstein, Podolsky, Rosen, and Shannon," quant-ph/0310010, 2003.

Petersen, A.

2005 "The Philosophy of Niels Bohr," in *Niels Bohr: A Centenary Volume*, eds. A. P. French and P. J. Kennedy. Cambridge, Mass.: Harvard University Press, 1985.

Plotnitsky, A.

1994 *Complementarity: Anti-Epistemology After Bohr and Derrida*, Durham, NC: Duke University Press, 1994.

2002 *The Knowable and the Unknowable: Modern Science, Nonclassical Thought, and the "Two Cultures,"* Ann Arbor, MI: University of Michigan Press, 2002.

2003 "Mysteries without Mysticism and Correlations without Correlata: On Quantum Knowledge and Knowledge in General," *Foundations of Physics* 33 (11) (2003), 1649-1689.

Plotnitsky, A. and Reed, D.

2001 "Discourse, Mathematics, Demonstration, and Science in Galileo's *Discourses Concerning Two New Sciences*," *Configurations* 9 (2001), 37-64.

Popper, K.

1982 *Quantum Theory and the Schism in Physics*, Totowa, NJ: Rowan and Littlefield, 1982.

Readhead, M.

1988 "A Philosopher Looks at Quantum Field Theory," in Brown, H. R. and Harré, R. *Philosophical Foundations of Quantum Field Theory*," Oxford: Clarendon, 1988, 9-23.

Reichenbach, H.

1956 *The Direction of Time*, Los Angeles: University of California Press, 1956.

Rovelli, C.

1996 "Relational Quantum Mechanics," *International Journal of Theoretical Physics* 35 (1996), 1637-1678.

Schilpp, A. P., ed.

1949 *Albert Einstein: Philosopher-Scientist*, New York: Tudor, 1949.

Schrödinger, E.

1935 "The Present Situation in Quantum Mechanics," in Wheeler, J. A. and Zurek, W. H. eds., *Quantum Theory and Measurement*, Princeton, NJ: Princeton University Press, 1983, 152-167.

Schweber, S. S.

1994 *QED and the Men Who Made It: Dyson, Feynman, Schwinger, and Tomonaga*, Princeton, NJ: Princeton University Press, 1994.

Stapp, H. P.

1987 "Quantum Nonlocality and the Description of Nature," in Cushing, J. T. and McMullin, E., eds., *Philosophical Consequences of Quantum Theory: Reflections on Bell's Theorem*, Notre Dame, IN: University of Notre Dame Press, 1989, 154-174.

1997 "Nonlocal Character of Quantum Theory," *American Journal of Physics* 65 (1997), 300-304.

Teller, P.

1995 *An Interpretive Introduction to Quantum Field Theory*, Princeton, NJ: Princeton University Press, 1995.

Ulfbeck, O. and Bohr, A.

2001 "Genuine Fortuitousness: Where Did That Click Come From?" *Foundations of Physics* 31 (5) (2001), 757-74.

Van Fraassen, B. C.

1991 *Quantum Mechanics: An Empiricist View*, Oxford: Clarendon, 1991.

Van der Waerden, B. L., ed.

1968 *Sources of Quantum Mechanics*, Toronto: Dover, 1968.

Von Neumann, J.

1961-1963 *Methods in Physical Sciences*, in *Collected Works of John von Neumann*, vol. 6 (Oxford: Pergamon Press, 1961-63).

1983 *Mathematical Foundations of Quantum Mechanics*, trans. Beyer, R.T., Princeton, NJ: Princeton University Press, 1983.

Weinberg, S.

2005 *The Quantum Theory of Fields, Volume 1: Foundations*, Cambridge: Cambridge University Press, 2005.

Weyl, H.

1918 *The Continuum: A Critical Examination of the Foundation of Analysis*, trans. Pollard, S. and Bole, T., New York: Dover, rpt. 1994.

1924 *Space Time Matter,* trans. Brose, H. L., New York; Dover, rpt. 1952.

Wheeler, J. A.

1983 "Law without Law," in Wheeler, J. A. and Zurek, W. H. *Quantum Theory and Measurement*, Princeton, NJ.: Princeton University Press 1983, 182-213.

1990 "Information, Physics, Quantum: The Search for Links," in Zurek, W. H. ed., *Complexity, Entropy and the Physics of Information*, Redwood, CA: Addison-Wesley, 1990, 3-28.

Wheeler, J. A. and Zurek, W. H.

1983 *Quantum Theory and Measurement*, Princeton, NJ: Princeton University Press, 1983.

Wilczek, F.

2005 "In Search of Symmetry Lost," *Nature* 423 (2005), 239-247.

Wittgenstein, L.

1985 *Tractatus Logico-Philosophicus*, trans. C. K. Ogden, Routledge: London, 1985.

Zee, A.

2003 *Quantum Field Theory in a Nutshell*, Princeton: Princeton University Press, 2003.

Zeilinger, A., Weith, G., Jennewein, T., and Aspelmeyer, M.

2005 "Happy Centenary, Photon," *Nature* 433 (2005), 230-238.

Zurek, W. H.

2003 "Decoherence, Einselection and the Quantum Qrigin of the Classical," *Review of Modern Physics* 75 (3) (2003), 715-775.

Name Index

Subject Index

For certain terms, such as "uncertainty relations," the pages where the key discussions of these terms occur are listed under the heading of "*key discussions.*" Due to the frequency of their occurrence several important terms, such as "quantum mechanics," "quantum objects," and "quantum processes," are only cross-indexed (e.g., "formalism, quantum-mechanical").

Fundamental Theories of Physics

Series Editor: Alwyn van der Merwe, University of Denver, USA

1. M. Sachs: *General Relativity and Matter.* A Spinor Field Theory from Fermis to Light-Years. With a Foreword by C. Kilmister. 1982 ISBN 90-277-1381-2
2. G.H. Duffey: *A Development of Quantum Mechanics.* Based on Symmetry Considerations. 1985 ISBN 90-277-1587-4
3. S. Diner, D. Fargue, G. Lochak and F. Selleri (eds.): *The Wave-Particle Dualism.* A Tribute to Louis de Broglie on his 90th Birthday. 1984 ISBN 90-277-1664-1
4. E. Prugovečki: *Stochastic Quantum Mechanics and Quantum Spacetime.* A Consistent Unification of Relativity and Quantum Theory based on Stochastic Spaces. 1984; 2nd printing 1986 ISBN 90-277-1617-X
5. D. Hestenes and G. Sobczyk: *Clifford Algebra to Geometric Calculus.* A Unified Language for Mathematics and Physics. 1984 ISBN 90-277-1673-0; Pb (1987) 90-277-2561-6
6. P. Exner: *Open Quantum Systems and Feynman Integrals.* 1985 ISBN 90-277-1678-1
7. L. Mayants: *The Enigma of Probability and Physics.* 1984 ISBN 90-277-1674-9
8. E. Tocaci: *Relativistic Mechanics, Time and Inertia.* Translated from Romanian. Edited and with a Foreword by C.W. Kilmister. 1985 ISBN 90-277-1769-9
9. B. Bertotti, F. de Felice and A. Pascolini (eds.): *General Relativity and Gravitation.* Proceedings of the 10th International Conference (Padova, Italy, 1983). 1984 ISBN 90-277-1819-9
10. G. Tarozzi and A. van der Merwe (eds.): *Open Questions in Quantum Physics.* 1985 ISBN 90-277-1853-9
11. J.V. Narlikar and T. Padmanabhan: *Gravity, Gauge Theories and Quantum Cosmology.* 1986 ISBN 90-277-1948-9
12. G.S. Asanov: *Finsler Geometry, Relativity and Gauge Theories.* 1985 ISBN 90-277-1960-8
13. K. Namsrai: *Nonlocal Quantum Field Theory and Stochastic Quantum Mechanics.* 1986 ISBN 90-277-2001-0
14. C. Ray Smith and W.T. Grandy, Jr. (eds.): *Maximum-Entropy and Bayesian Methods in Inverse Problems.* Proceedings of the 1st and 2nd International Workshop (Laramie, Wyoming, USA). 1985 ISBN 90-277-2074-6
15. D. Hestenes: *New Foundations for Classical Mechanics.* 1986 ISBN 90-277-2090-8; Pb (1987) 90-277-2526-8
16. S.J. Prokhovnik: *Light in Einstein's Universe.* The Role of Energy in Cosmology and Relativity. 1985 ISBN 90-277-2093-2
17. Y.S. Kim and M.E. Noz: *Theory and Applications of the Poincaré Group.* 1986 ISBN 90-277-2141-6
18. M. Sachs: *Quantum Mechanics from General Relativity.* An Approximation for a Theory of Inertia. 1986 ISBN 90-277-2247-1
19. W.T. Grandy, Jr.: *Foundations of Statistical Mechanics.* Vol. I: *Equilibrium Theory.* 1987 ISBN 90-277-2489-X
20. H.-H von Borzeszkowski and H.-J. Treder: *The Meaning of Quantum Gravity.* 1988 ISBN 90-277-2518-7
21. C. Ray Smith and G.J. Erickson (eds.): *Maximum-Entropy and Bayesian Spectral Analysis and Estimation Problems.* Proceedings of the 3rd International Workshop (Laramie, Wyoming, USA, 1983). 1987 ISBN 90-277-2579-9
22. A.O. Barut and A. van der Merwe (eds.): *Selected Scientific Papers of Alfred Landé.* [*1888-1975*]. 1988 ISBN 90-277-2594-2

Fundamental Theories of Physics

23. W.T. Grandy, Jr.: *Foundations of Statistical Mechanics.* Vol. II: *Nonequilibrium Phenomena.* 1988 ISBN 90-277-2649-3
24. E.I. Bitsakis and C.A. Nicolaides (eds.): *The Concept of Probability.* Proceedings of the Delphi Conference (Delphi, Greece, 1987). 1989 ISBN 90-277-2679-5
25. A. van der Merwe, F. Selleri and G. Tarozzi (eds.): *Microphysical Reality and Quantum Formalism, Vol. 1.* Proceedings of the International Conference (Urbino, Italy, 1985). 1988 ISBN 90-277-2683-3
26. A. van der Merwe, F. Selleri and G. Tarozzi (eds.): *Microphysical Reality and Quantum Formalism, Vol. 2.* Proceedings of the International Conference (Urbino, Italy, 1985). 1988 ISBN 90-277-2684-1
27. I.D. Novikov and V.P. Frolov: *Physics of Black Holes.* 1989 ISBN 90-277-2685-X
28. G. Tarozzi and A. van der Merwe (eds.): *The Nature of Quantum Paradoxes.* Italian Studies in the Foundations and Philosophy of Modern Physics. 1988 ISBN 90-277-2703-1
29. B.R. Iyer, N. Mukunda and C.V. Vishveshwara (eds.): *Gravitation, Gauge Theories and the Early Universe.* 1989 ISBN 90-277-2710-4
30. H. Mark and L. Wood (eds.): *Energy in Physics, War and Peace.* A Festschrift celebrating Edward Teller's 80th Birthday. 1988 ISBN 90-277-2775-9
31. G.J. Erickson and C.R. Smith (eds.): *Maximum-Entropy and Bayesian Methods in Science and Engineering.* Vol. I: *Foundations.* 1988 ISBN 90-277-2793-7
32. G.J. Erickson and C.R. Smith (eds.): *Maximum-Entropy and Bayesian Methods in Science and Engineering.* Vol. II: *Applications.* 1988 ISBN 90-277-2794-5
33. M.E. Noz and Y.S. Kim (eds.): *Special Relativity and Quantum Theory.* A Collection of Papers on the Poincaré Group. 1988 ISBN 90-277-2799-6
34. I.Yu. Kobzarev and Yu.I. Manin: *Elementary Particles. Mathematics, Physics and Philosophy.* 1989 ISBN 0-7923-0098-X
35. F. Selleri: *Quantum Paradoxes and Physical Reality.* 1990 ISBN 0-7923-0253-2
36. J. Skilling (ed.): *Maximum-Entropy and Bayesian Methods.* Proceedings of the 8th International Workshop (Cambridge, UK, 1988). 1989 ISBN 0-7923-0224-9
37. M. Kafatos (ed.): *Bell's Theorem, Quantum Theory and Conceptions of the Universe.* 1989 ISBN 0-7923-0496-9
38. Yu.A. Izyumov and V.N. Syromyatnikov: *Phase Transitions and Crystal Symmetry.* 1990 ISBN 0-7923-0542-6
39. P.F. Fougère (ed.): *Maximum-Entropy and Bayesian Methods.* Proceedings of the 9th International Workshop (Dartmouth, Massachusetts, USA, 1989). 1990 ISBN 0-7923-0928-6
40. L. de Broglie: *Heisenberg's Uncertainties and the Probabilistic Interpretation of Wave Mechanics.* With Critical Notes of the Author. 1990 ISBN 0-7923-0929-4
41. W.T. Grandy, Jr.: *Relativistic Quantum Mechanics of Leptons and Fields.* 1991 ISBN 0-7923-1049-7
42. Yu.L. Klimontovich: *Turbulent Motion and the Structure of Chaos.* A New Approach to the Statistical Theory of Open Systems. 1991 ISBN 0-7923-1114-0
43. W.T. Grandy, Jr. and L.H. Schick (eds.): *Maximum-Entropy and Bayesian Methods.* Proceedings of the 10th International Workshop (Laramie, Wyoming, USA, 1990). 1991 ISBN 0-7923-1140-X
44. P. Pták and S. Pulmannová: *Orthomodular Structures as Quantum Logics.* Intrinsic Properties, State Space and Probabilistic Topics. 1991 ISBN 0-7923-1207-4
45. D. Hestenes and A. Weingartshofer (eds.): *The Electron.* New Theory and Experiment. 1991 ISBN 0-7923-1356-9

Fundamental Theories of Physics

46. P.P.J.M. Schram: *Kinetic Theory of Gases and Plasmas.* 1991 ISBN 0-7923-1392-5
47. A. Micali, R. Boudet and J. Helmstetter (eds.): *Clifford Algebras and their Applications in Mathematical Physics.* 1992 ISBN 0-7923-1623-1
48. E. Prugovečki: *Quantum Geometry.* A Framework for Quantum General Relativity. 1992 ISBN 0-7923-1640-1
49. M.H. Mac Gregor: *The Enigmatic Electron.* 1992 ISBN 0-7923-1982-6
50. C.R. Smith, G.J. Erickson and P.O. Neudorfer (eds.): *Maximum Entropy and Bayesian Methods.* Proceedings of the 11th International Workshop (Seattle, 1991). 1993 ISBN 0-7923-2031-X
51. D.J. Hoekzema: *The Quantum Labyrinth.* 1993 ISBN 0-7923-2066-2
52. Z. Oziewicz, B. Jancewicz and A. Borowiec (eds.): *Spinors, Twistors, Clifford Algebras and Quantum Deformations.* Proceedings of the Second Max Born Symposium (Wrocław, Poland, 1992). 1993 ISBN 0-7923-2251-7
53. A. Mohammad-Djafari and G. Demoment (eds.): *Maximum Entropy and Bayesian Methods.* Proceedings of the 12th International Workshop (Paris, France, 1992). 1993 ISBN 0-7923-2280-0
54. M. Riesz: *Clifford Numbers and Spinors* with Riesz' Private Lectures to E. Folke Bolinder and a Historical Review by Pertti Lounesto. E.F. Bolinder and P. Lounesto (eds.). 1993 ISBN 0-7923-2299-1
55. F. Brackx, R. Delanghe and H. Serras (eds.): *Clifford Algebras and their Applications in Mathematical Physics.* Proceedings of the Third Conference (Deinze, 1993) 1993 ISBN 0-7923-2347-5
56. J.R. Fanchi: *Parametrized Relativistic Quantum Theory.* 1993 ISBN 0-7923-2376-9
57. A. Peres: *Quantum Theory: Concepts and Methods.* 1993 ISBN 0-7923-2549-4
58. P.L. Antonelli, R.S. Ingarden and M. Matsumoto: *The Theory of Sprays and Finsler Spaces with Applications in Physics and Biology.* 1993 ISBN 0-7923-2577-X
59. R. Miron and M. Anastasiei: *The Geometry of Lagrange Spaces: Theory and Applications.* 1994 ISBN 0-7923-2591-5
60. G. Adomian: *Solving Frontier Problems of Physics: The Decomposition Method.* 1994 ISBN 0-7923-2644-X
61. B.S. Kerner and V.V. Osipov: *Autosolitons.* A New Approach to Problems of Self-Organization and Turbulence. 1994 ISBN 0-7923-2816-7
62. G.R. Heidbreder (ed.): *Maximum Entropy and Bayesian Methods.* Proceedings of the 13th International Workshop (Santa Barbara, USA, 1993) 1996 ISBN 0-7923-2851-5
63. J. Peřina, Z. Hradil and B. Jurčo: *Quantum Optics and Fundamentals of Physics.* 1994 ISBN 0-7923-3000-5
64. M. Evans and J.-P. Vigier: *The Enigmatic Photon.* Volume 1: The Field $\boldsymbol{B}^{(3)}$. 1994 ISBN 0-7923-3049-8
65. C.K. Raju: *Time: Towards a Constistent Theory.* 1994 ISBN 0-7923-3103-6
66. A.K.T. Assis: *Weber's Electrodynamics.* 1994 ISBN 0-7923-3137-0
67. Yu. L. Klimontovich: *Statistical Theory of Open Systems.* Volume 1: A Unified Approach to Kinetic Description of Processes in Active Systems. 1995 ISBN 0-7923-3199-0; Pb: ISBN 0-7923-3242-3
68. M. Evans and J.-P. Vigier: *The Enigmatic Photon.* Volume 2: Non-Abelian Electrodynamics. 1995 ISBN 0-7923-3288-1
69. G. Esposito: *Complex General Relativity.* 1995 ISBN 0-7923-3340-3

Fundamental Theories of Physics

70. J. Skilling and S. Sibisi (eds.): *Maximum Entropy and Bayesian Methods.* Proceedings of the Fourteenth International Workshop on Maximum Entropy and Bayesian Methods. 1996 ISBN 0-7923-3452-3
71. C. Garola and A. Rossi (eds.): *The Foundations of Quantum Mechanics Historical Analysis and Open Questions.* 1995 ISBN 0-7923-3480-9
72. A. Peres: *Quantum Theory: Concepts and Methods.* 1995 (see for hardback edition, Vol. 57) ISBN Pb 0-7923-3632-1
73. M. Ferrero and A. van der Merwe (eds.): *Fundamental Problems in Quantum Physics.* 1995 ISBN 0-7923-3670-4
74. F.E. Schroeck, Jr.: *Quantum Mechanics on Phase Space.* 1996 ISBN 0-7923-3794-8
75. L. de la Peña and A.M. Cetto: *The Quantum Dice.* An Introduction to Stochastic Electrodynamics. 1996 ISBN 0-7923-3818-9
76. P.L. Antonelli and R. Miron (eds.): *Lagrange and Finsler Geometry.* Applications to Physics and Biology. 1996 ISBN 0-7923-3873-1
77. M.W. Evans, J.-P. Vigier, S. Roy and S. Jeffers: *The Enigmatic Photon.* Volume 3: Theory and Practice of the $\boldsymbol{B}^{(3)}$ Field. 1996 ISBN 0-7923-4044-2
78. W.G.V. Rosser: *Interpretation of Classical Electromagnetism.* 1996 ISBN 0-7923-4187-2
79. K.M. Hanson and R.N. Silver (eds.): *Maximum Entropy and Bayesian Methods.* 1996 ISBN 0-7923-4311-5
80. S. Jeffers, S. Roy, J.-P. Vigier and G. Hunter (eds.): *The Present Status of the Quantum Theory of Light.* Proceedings of a Symposium in Honour of Jean-Pierre Vigier. 1997 ISBN 0-7923-4337-9
81. M. Ferrero and A. van der Merwe (eds.): *New Developments on Fundamental Problems in Quantum Physics.* 1997 ISBN 0-7923-4374-3
82. R. Miron: *The Geometry of Higher-Order Lagrange Spaces.* Applications to Mechanics and Physics. 1997 ISBN 0-7923-4393-X
83. T. Hakioğlu and A.S. Shumovsky (eds.): *Quantum Optics and the Spectroscopy of Solids.* Concepts and Advances. 1997 ISBN 0-7923-4414-6
84. A. Sitenko and V. Tartakovskii: *Theory of Nucleus.* Nuclear Structure and Nuclear Interaction. 1997 ISBN 0-7923-4423-5
85. G. Esposito, A.Yu. Kamenshchik and G. Pollifrone: *Euclidean Quantum Gravity on Manifolds with Boundary.* 1997 ISBN 0-7923-4472-3
86. R.S. Ingarden, A. Kossakowski and M. Ohya: *Information Dynamics and Open Systems.* Classical and Quantum Approach. 1997 ISBN 0-7923-4473-1
87. K. Nakamura: *Quantum versus Chaos.* Questions Emerging from Mesoscopic Cosmos. 1997 ISBN 0-7923-4557-6
88. B.R. Iyer and C.V. Vishveshwara (eds.): *Geometry, Fields and Cosmology.* Techniques and Applications. 1997 ISBN 0-7923-4725-0
89. G.A. Martynov: *Classical Statistical Mechanics.* 1997 ISBN 0-7923-4774-9
90. M.W. Evans, J.-P. Vigier, S. Roy and G. Hunter (eds.): *The Enigmatic Photon.* Volume 4: New Directions. 1998 ISBN 0-7923-4826-5
91. M. Rédei: *Quantum Logic in Algebraic Approach.* 1998 ISBN 0-7923-4903-2
92. S. Roy: *Statistical Geometry and Applications to Microphysics and Cosmology.* 1998 ISBN 0-7923-4907-5
93. B.C. Eu: *Nonequilibrium Statistical Mechanics.* Ensembled Method. 1998 ISBN 0-7923-4980-6

Fundamental Theories of Physics

94. V. Dietrich, K. Habetha and G. Jank (eds.): *Clifford Algebras and Their Application in Mathematical Physics.* Aachen 1996. 1998 ISBN 0-7923-5037-5
95. J.P. Blaizot, X. Campi and M. Ploszajczak (eds.): *Nuclear Matter in Different Phases and Transitions.* 1999 ISBN 0-7923-5660-8
96. V.P. Frolov and I.D. Novikov: *Black Hole Physics.* Basic Concepts and New Developments. 1998 ISBN 0-7923-5145-2; Pb 0-7923-5146
97. G. Hunter, S. Jeffers and J-P. Vigier (eds.): *Causality and Locality in Modern Physics.* 1998 ISBN 0-7923-5227-0
98. G.J. Erickson, J.T. Rychert and C.R. Smith (eds.): *Maximum Entropy and Bayesian Methods.* 1998 ISBN 0-7923-5047-2
99. D. Hestenes: *New Foundations for Classical Mechanics (Second Edition).* 1999 ISBN 0-7923-5302-1; Pb ISBN 0-7923-5514-8
100. B.R. Iyer and B. Bhawal (eds.): *Black Holes, Gravitational Radiation and the Universe.* Essays in Honor of C. V. Vishveshwara. 1999 ISBN 0-7923-5308-0
101. P.L. Antonelli and T.J. Zastawniak: *Fundamentals of Finslerian Diffusion with Applications.* 1998 ISBN 0-7923-5511-3
102. H. Atmanspacher, A. Amann and U. Müller-Herold: *On Quanta, Mind and Matter Hans Primas in Context.* 1999 ISBN 0-7923-5696-9
103. M.A. Trump and W.C. Schieve: *Classical Relativistic Many-Body Dynamics.* 1999 ISBN 0-7923-5737-X
104. A.I. Maimistov and A.M. Basharov: *Nonlinear Optical Waves.* 1999 ISBN 0-7923-5752-3
105. W. von der Linden, V. Dose, R. Fischer and R. Preuss (eds.): *Maximum Entropy and Bayesian Methods Garching, Germany 1998.* 1999 ISBN 0-7923-5766-3
106. M.W. Evans: *The Enigmatic Photon Volume 5: O(3) Electrodynamics.* 1999 ISBN 0-7923-5792-2
107. G.N. Afanasiev: *Topological Effects in Quantum Mecvhanics.* 1999 ISBN 0-7923-5800-7
108. V. Devanathan: *Angular Momentum Techniques in Quantum Mechanics.* 1999 ISBN 0-7923-5866-X
109. P.L. Antonelli (ed.): *Finslerian Geometries A Meeting of Minds.* 1999 ISBN 0-7923-6115-6
110. M.B. Mensky: *Quantum Measurements and Decoherence Models and Phenomenology.* 2000 ISBN 0-7923-6227-6
111. B. Coecke, D. Moore and A. Wilce (eds.): *Current Research in Operation Quantum Logic.* Algebras, Categories, Languages. 2000 ISBN 0-7923-6258-6
112. G. Jumarie: *Maximum Entropy, Information Without Probability and Complex Fractals.* Classical and Quantum Approach. 2000 ISBN 0-7923-6330-2
113. B. Fain: *Irreversibilities in Quantum Mechanics.* 2000 ISBN 0-7923-6581-X
114. T. Borne, G. Lochak and H. Stumpf: *Nonperturbative Quantum Field Theory and the Structure of Matter.* 2001 ISBN 0-7923-6803-7
115. J. Keller: *Theory of the Electron.* A Theory of Matter from START. 2001 ISBN 0-7923-6819-3
116. M. Rivas: *Kinematical Theory of Spinning Particles.* Classical and Quantum Mechanical Formalism of Elementary Particles. 2001 ISBN 0-7923-6824-X
117. A.A. Ungar: *Beyond the Einstein Addition Law and its Gyroscopic Thomas Precession.* The Theory of Gyrogroups and Gyrovector Spaces. 2001 ISBN 0-7923-6909-2
118. R. Miron, D. Hrimiuc, H. Shimada and S.V. Sabau: *The Geometry of Hamilton and Lagrange Spaces.* 2001 ISBN 0-7923-6926-2

Fundamental Theories of Physics

119. M. Pavšič: *The Landscape of Theoretical Physics: A Global View*. From Point Particles to the Brane World and Beyond in Search of a Unifying Principle. 2001 ISBN 0-7923-7006-6
120. R.M. Santilli: *Foundations of Hadronic Chemistry*. With Applications to New Clean Energies and Fuels. 2001 ISBN 1-4020-0087-1
121. S. Fujita and S. Godoy: *Theory of High Temperature Superconductivity*. 2001 ISBN 1-4020-0149-5
122. R. Luzzi, A.R. Vasconcellos and J. Galvão Ramos: *Predictive Statitical Mechanics*. A Nonequilibrium Ensemble Formalism. 2002 ISBN 1-4020-0482-6
123. V.V. Kulish: *Hierarchical Methods*. Hierarchy and Hierarchical Asymptotic Methods in Electrodynamics, Volume 1. 2002 ISBN 1-4020-0757-4; Set: 1-4020-0758-2
124. B.C. Eu: *Generalized Thermodynamics*. Thermodynamics of Irreversible Processes and Generalized Hydrodynamics. 2002 ISBN 1-4020-0788-4
125. A. Mourachkine: *High-Temperature Superconductivity in Cuprates*. The Nonlinear Mechanism and Tunneling Measurements. 2002 ISBN 1-4020-0810-4
126. R.L. Amoroso, G. Hunter, M. Kafatos and J.-P. Vigier (eds.): *Gravitation and Cosmology: From the Hubble Radius to the Planck Scale*. Proceedings of a Symposium in Honour of the 80th Birthday of Jean-Pierre Vigier. 2002 ISBN 1-4020-0885-6
127. W.M. de Muynck: *Foundations of Quantum Mechanics, an Empiricist Approach*. 2002 ISBN 1-4020-0932-1
128. V.V. Kulish: *Hierarchical Methods*. Undulative Electrodynamical Systems, Volume 2. 2002 ISBN 1-4020-0968-2; Set: 1-4020-0758-2
129. M. Mugur-Schächter and A. van der Merwe (eds.): *Quantum Mechanics, Mathematics, Cognition and Action*. Proposals for a Formalized Epistemology. 2002 ISBN 1-4020-1120-2
130. P. Bandyopadhyay: *Geometry, Topology and Quantum Field Theory*. 2003 ISBN 1-4020-1414-7
131. V. Garzó and A. Santos: *Kinetic Theory of Gases in Shear Flows*. Nonlinear Transport. 2003 ISBN 1-4020-1436-8
132. R. Miron: *The Geometry of Higher-Order Hamilton Spaces*. Applications to Hamiltonian Mechanics. 2003 ISBN 1-4020-1574-7
133. S. Esposito, E. Majorana Jr., A. van der Merwe and E. Recami (eds.): *Ettore Majorana: Notes on Theoretical Physics*. 2003 ISBN 1-4020-1649-2
134. J. Hamhalter. *Quantum Measure Theory*. 2003 ISBN 1-4020-1714-6
135. G. Rizzi and M.L. Ruggiero: *Relativity in Rotating Frames*. Relativistic Physics in Rotating Reference Frames. 2004 ISBN 1-4020-1805-3
136. L. Kantorovich: *Quantum Theory of the Solid State: an Introduction*. 2004 ISBN 1-4020-1821-5
137. A. Ghatak and S. Lokanathan: *Quantum Mechanics: Theory and Applications*. 2004 ISBN 1-4020-1850-9
138. A. Khrennikov: *Information Dynamics in Cognitive, Psychological, Social, and Anomalous Phenomena*. 2004 ISBN 1-4020-1868-1
139. V. Faraoni: *Cosmology in Scalar-Tensor Gravity*. 2004 ISBN 1-4020-1988-2
140. P.P. Teodorescu and N.-A. P. Nicorovici: *Applications of the Theory of Groups in Mechanics and Physics*. 2004 ISBN 1-4020-2046-5
141. G. Munteanu: *Complex Spaces in Finsler, Lagrange and Hamilton Geometries*. 2004 ISBN 1-4020-2205-0

Fundamental Theories of Physics

142. G.N. Afanasiev: *Vavilov-Cherenkov and Synchrotron Radiation*. Foundations and Applications. 2004 ISBN 1-4020-2410-X
143. L. Munteanu and S. Donescu: *Introduction to Soliton Theory: Applications to Mechanics*. 2004 ISBN 1-4020-2576-9
144. M.Yu. Khlopov and S.G. Rubin: *Cosmological Pattern of Microphysics in the Inflationary Universe*. 2004 ISBN 1-4020-2649-8
145. J. Vanderlinde: *Classical Electromagnetic Theory*. 2004 ISBN 1-4020-2699-4
146. V. Čápek and D.P. Sheehan: *Challenges to the Second Law of Thermodynamics*. Theory and Experiment. 2005 ISBN 1-4020-3015-0
147. B.G. Sidharth: *The Universe of Fluctuations*. The Architecture of Spacetime and the Universe. 2005 ISBN 1-4020-3785-6
148. R.W. Carroll: *Fluctuations, Information, Gravity and the Quantum Potential*. 2005 ISBN 1-4020-4003-2
149. B.G. Sidharth: *A Century of Ideas*. Personal Perspectives from a Selection of the Greatest Minds of the Twentieth Century. Planned 2006. ISBN 1-4020-4359-7
150. S.H. Dong: *Factorization Method in Quantum Mechanics*. Planned 2006. ISBN to be announced
151. R.M. Santilli: *Isodual Theory of Antimatter with applications to Antigravity, Grand Unification and Cosmology*. 2006 ISBN 1-4020-4517-4
152. A. Plotnitsky: *Reading Bohr: Physics and Philosophy*. 2006 ISBN 1-4020-5253-7